Introduction to Algebra and Geometry

Doug Gardner
Rogue Community College

arranged by Tim Merzenich
Chemeketa Community College

A Chemeketa Press Reprint

Introduction to Algebra and Geometry

A Chemeketa Press Reprint

ISBN: 978-1-943536-03-0

The contents of this book come from Chapters 1-3 of *Applied Algebra I* and chapter 1 of *Applied Algebra II,* by Doug Gardner, Rogue Community College. These books were written with funding from National Science Foundation grant, award numbers 1002822 and 1304307.

All materials © Doug Gardner. Used by permission of the author.

This book was designed and produced by the students and staff of the Chemeketa Visual Communications Program.

Chemeketa Press
Managing Editor: Steve Richardson
Art Director: Kristen MacDonald
Layout: Terra Hyle
Cover Design: Cierra Maher
Cover Photo: "Brooklyn Bridge," © 2012 by Cierra Maher. Used by permission of the artist.

Printed in the United States of America.

Contents

Legend of Assigment Icons

Manufacturing   Electronics   Construction   Automotive   Welding   Health

About this Book:

This book is designed to be read. Some humor and stories of my personal journey in understanding and applying this beautiful language are included to keep it interesting. There are <u>no</u> shortcuts, however! If you are to understand the material deeply enough to apply it to all the rich problems that life presents, you will need to add a considerable amount of your own thought and dedication to the process. The biggest problem math students have is attempting to memorize steps to solve problems. This guarantees that you will not remember it well enough, or know it deeply enough, to apply it. All mathematics can be <u>understood</u> by any student willing to invest the time to ask questions, think and scribble (and sometimes throw things). Your goal should be to "own" the material rather than merely "borrow" it. Owning it means that you carry the knowledge with you, like a carpenter carries his or her tools, ready to use them as needed. Ownership, however, comes at a cost and the amount you must pay is dependent on your natural ability, interest and the time you have to study. Remember, if it was as easy as digging a ditch, everyone could do it and you would not get paid extra for knowing it.

After a brief introduction, each section will offer examples to introduce new concepts and show how they can be applied. The examples are not intended to provide a step-by-step procedure for you to copy when you solve the problems in the assignment. In fact, you will find that the assigned problems may not even resemble the examples. Real-life problems, such as those that arise in the workplace, are so varied that you cannot expect to follow a memorized set of steps to solve them.

Algebra is, ultimately, a useful method for expressing relationships in the world that people use every day. Every section of this book is devoted to exposing you to, and increasing your skill with, formulas (numerical relationships). Of course there are relationships in the world that you may find more interesting, but you are probably reading this because you have already discovered that you cannot live on love alone.

Consider the common experience of making a car payment. You understand that there is a relationship between your monthly car payment and the price of the car, the interest rate, and the time it takes to pay off the loan. If either the car's price or the interest rate go up, so will the monthly payment. On the other hand, if you increase the time period over which you will pay off the loan, the monthly payment will decrease. Without the algebraic formula, however, there would be no way to figure out what the monthly payment should be. The formula that expresses this relationship precisely is:

$$M = \frac{PRI^T}{12(I^T - 1)}$$

Where M = monthly payment, P = price of the car, R = annual percentage rate (APR), $I = 1 + \frac{R}{12}$ (the modified interest rate representing monthly growth), and T = time to pay off the loan in months.

Though this is a daunting formula, complete with multiplication, division, addition, subtraction, exponents and parenthesis, indeed, every operation that will be studied in this course, the meaning and order of operations can be understood by anyone willing to invest the time necessary to speak this most interesting and useful language called algebra.

Although you might not understand all the details until the end of the course, let's attempt to calculate a car payment. Don't worry if you find it too difficult right now. This should motivate your study of algebra since money is an arena of life that every person spends some time caged in, and algebra provides an elegant means of escape.

Example: Calculating a monthly payment

Suppose you purchase a used car for $8000, to be paid off in 48 months, at 6% APR (annual percentage rate).

Calculate the monthly payment (M) using the formula: $M = \dfrac{PRI^T}{12(I^T - 1)}$

Note that P = 8000, T = 48 and R = .06.

Solution:

$I = 1 + \dfrac{R}{12}$

First, find the value for I. To do this, enter 1 + .06 ÷ 12 in your calculator (you should get **1.005**).

I^T

Second, find the value for I^T. Since I = 1.005 and T = 48, we are finding 1.005 to the 48th power. The calculator entry will be 1.005 ^ 48 or 1.005 x^y 48, depending on the type of calculator that you have (you should get **1.27**, if you round your answer to 2 decimal places).

$M \approx \dfrac{8000*.06*1.27}{12(1.27-1)}$

Third, enter the numbers into the formula.

8000 x .06 x 1.27

Fourth, simplify the top (numerator) of the fraction (you should get **609.6**).

12(1.27 - 1)

Fifth, simplify the bottom (denominator) of the fraction. The quantity in the parentheses is equal to .27. Thus, you can enter 12 x .27 in your calculator to get the value of the denominator (you should get **3.24**).

609.6 ÷ 3.24

Finally, finish the division (you should get **$188.15**, when you round to 2 decimal places).

Final Answer: The monthly payment of **$188.15** would pay off the car including interest in 48 months or 4 years.

Note: a bank would more accurately get $187.88, since the 1.27 in the 2nd step should be done without rounding.

Although the monthly payment formula is complex, you may not understand the details of the order of operations, or how to use your calculator skillfully, appreciate that algebra is a very practical tool.

This course will focus on the study of algebraic relationships such as may be directly applied to careers in Electronics, Manufacturing, Construction, Diesel, Automotive and Welding. Admittedly, you will face challenges attempting to scale this mountain of understanding. It is not without reason that I see a pained expression when I tell people that I teach math for a living. Climbing a mountain is difficult for anyone, but if you address yourself to the task, you will find your stride and the view from the top will justify the energy you expend. Let's begin in chapter 1 with an equipment check to make sure you have the fuel and fitness to hit the trail with a shot at the summit.

Chapter 1:
Tools of Algebra

Using algebraic formulas requires skill with all types of numbers (negatives, fractions and decimals), an understanding of measurements for lengths and angles, familiarity with units (such as feet, mm^2, in^3), and a firm grasp of the order of operations.   We are at the trail head and this is an equipment check.   You should notice that you already have much of what you will need.   This chapter will refresh your memory, deepen your grasp of the basics, and introduce a few new concepts that you will find useful on the trail.

1.1: Operations with Real Numbers

The study of Algebra must begin with a review of the basic operations (+, -, x, ÷) on the numbers that show up in the real world (negatives, fractions, and decimals). Stirred together, they are appropriately called real numbers (such as 5, -3.62, -⅝). Recall the facts for negative numbers and fractions:

Negative Number Facts for Multiplication:	Fraction Facts:
1. (+)(+) = +	1. $\dfrac{a}{b} + \dfrac{c}{d} = \dfrac{ad+bc}{bd}$
2. (+)(-) = -	2. $\dfrac{a}{b} - \dfrac{c}{d} = \dfrac{ad-bc}{bd}$
3. (-)(+) = -	3. $\dfrac{a}{b} \times \dfrac{c}{d} = \dfrac{ac}{bd}$
4. (-)(-) = +	4. $\dfrac{a}{b} \div \dfrac{c}{d} = \dfrac{a}{b} \times \dfrac{d}{c} = \dfrac{ad}{bc}$

Side note: -(-7) is quite logically +7 since -(-7) could represent taking away a debt of $7. Taking away debt is always positive.

Let's review multiplication with decimals first:

Example 1.1.1: Multiplying decimals with negatives

Simplify: -4.2 • (-6.3)

Solution:

Stop and think: The answer will be positive and it is always a good idea to estimate first: -4 • (-6) = 24 , so we can expect the exact answer to be larger than 24.

Final Answer: 26.46

Apply the negative number facts to an example with fractions:

Example 1.1.2: Dividing fractions with negatives

Simplify: $6\frac{2}{3} \div -5\frac{1}{4}$

Solution:

Stop and think: The answer will be negative and it is always a good idea to estimate first: 6 ÷ 5 will be a little larger than 1. We can ignore the sign during the work and just include it in the answer:

$6\frac{2}{3} \div 5\frac{1}{4}$

$\frac{20}{3} \div \frac{21}{4}$ $6 = \frac{18}{3}$ and $\frac{18}{3} + \frac{2}{3} = \frac{20}{3}$; $5 = \frac{20}{4}$ and $\frac{20}{4} + \frac{1}{4} = \frac{21}{4}$

$\frac{20}{3} \times \frac{4}{21}$ invert (flip) the second fraction then multiply

$\frac{80}{63}$ fractions are multiplied straight across

$1\frac{17}{63}$ $\frac{80}{63} = \frac{63}{63} + \frac{17}{63} = 1 + \frac{17}{63}$

Final Answer: $-1\frac{17}{63}$

Addition and subtraction with negative numbers is simple if you think of the sign as a symbol for direction. A football team could gain 12 yards (+12) or lose 12 yards (-12). The 12 tells us the distance traveled, while the sign tells us the direction. It is an unfortunate but unavoidable fact that all of life's experiences are not positive. A carpenter can cut a rafter that is off by ⅛ inch. +⅛ represents ⅛" too long and -⅛ represents ⅛" too short. Operations with negative numbers are easy to picture if you use a number line to represent direction. Moving right is positive, left is negative.

Examples:

1. $-5 - 4 = -9$: begin at -5 then move left 4 to -9
2. $-2 - (-7) = 5$: begin at -2 and move right 7 to 5 (right because a (-)(-) is a positive)

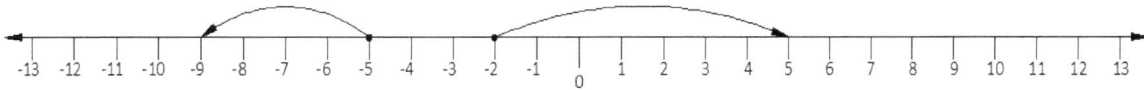

It is interesting to note that, despite the variety signs add to a problem involving a 6 and a 2,

a) $-6 + 2$ b) $-6 - 2$ c) $-2 + 6$ d) $-2 - 6$

e) $6 + 2$ f) $6 - 2$ g) $2 + 6$ h) $2 - 6$

the absolute value of the answer (the value of the number ignoring the sign) must either be 8 or 4 in every case. Study the illustration below:

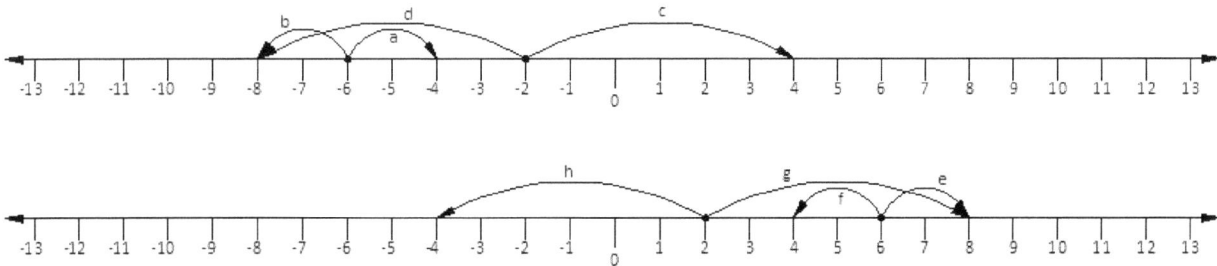

Subtracting and adding with negative numbers is as simple as beginning at the first number then moving the distance and direction specified by the second number. The process can be understood so that you do not need a list of rules and steps to follow for each different case.

Let's work an addition example:

Example 1.1.3: Adding negative numbers

Simplify: $-3 + (-7)$

Solution:
Start at -3, and move left 7.

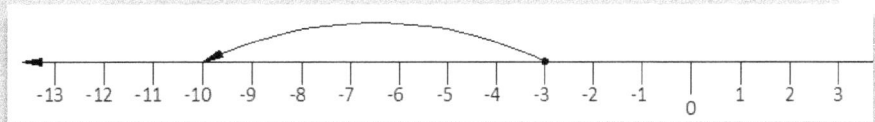

Final Answer: -10

Let's work a subtraction example:

Example 1.1.4: Subtracting negative numbers

Simplify: -8 − (-13)

Solution:

Start at -8, and move right 13.

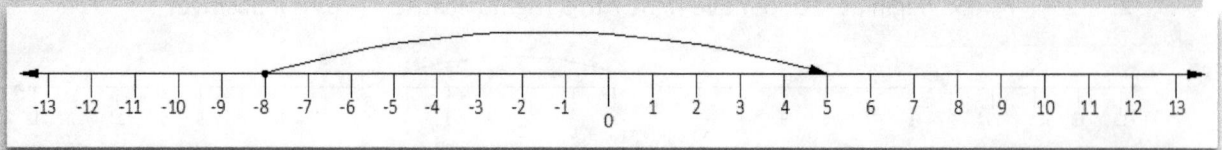

Final Answer: 5

Notice that in example 1.1.3 we ended up <u>adding</u> the 3 and 7 since we started by moving left to -3 then left again 7 more. In example 1.1.4 we <u>subtracted</u> the 13 and 8 since we started by moving left to -8 then 13 back to the right. We could proceed to a rule here but you won't need it if you have a picture of what is happening on the number line.

Let's extend our understanding to include decimals:

Example 1.1.5: Subtracting decimals with negatives

Simplify: -4.548 − 7.289

Solution:

Start at -4.548 and move left 7.289. You will add the numbers and be on the negative side of 0.

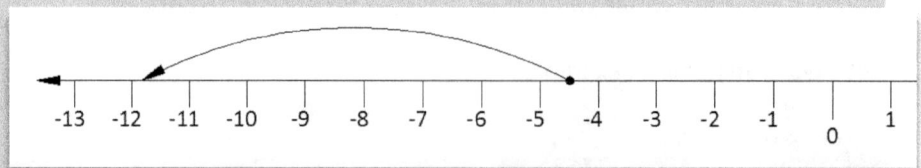

Final Answer: -11.837

Let's extend our understanding to include fractions:

Example 1.1.6: Adding fractions with negatives

Simplify: $4\frac{5}{8} + (-9\frac{3}{4})$

Solution:

Start at $4\frac{5}{8}$ and move left $9\frac{3}{4}$. You will subtract the numbers and be on the negative side of 0.

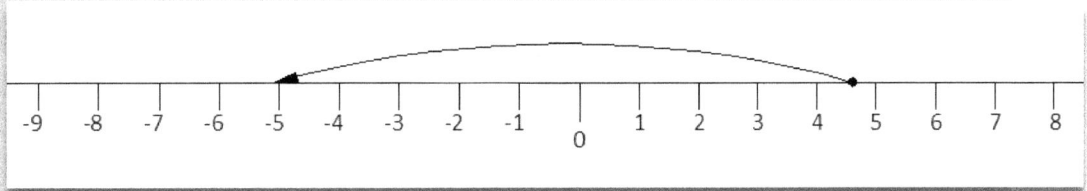

$9\frac{3}{4} - 4\frac{5}{8}$	set up the subtraction problem
$9\frac{6}{8} - 4\frac{5}{8}$	find a common denominator for 4 and 8
$5\frac{1}{8}$	simplify
$-5\frac{1}{8}$	include the negative sign since the number is on the negative side of zero

Final Answer: $-5\frac{1}{8}$

Section 1.1: Operations with Real Numbers

1. A carpenter needs a board that is $79\frac{5}{8}"$ long that she will cut from a board is $96\frac{1}{4}"$ long. If the saw blade has a $\frac{3}{16}"$ kerf blade (kerf is the thickness of wood that the blade removes with a cut), calculate the length of the left over piece.

2. Calculate the missing dimensions A, B, C, and D. Measurements are in inches.

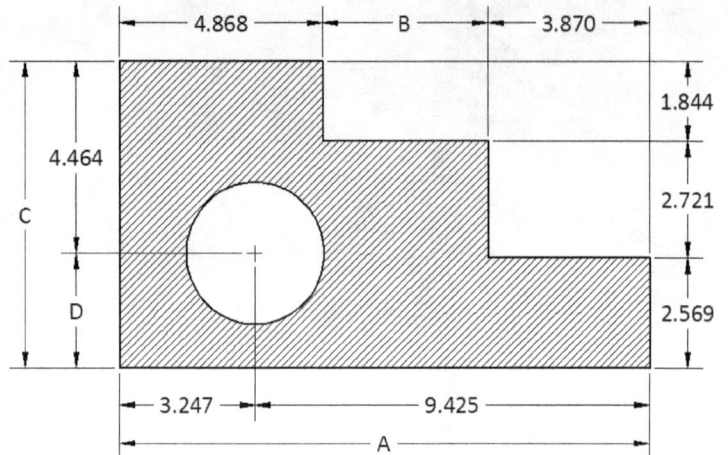

3. Calculate the length of the non-threaded portion of the bolt (S). Calculate the size of the overhang (H). Measurements are in inches.

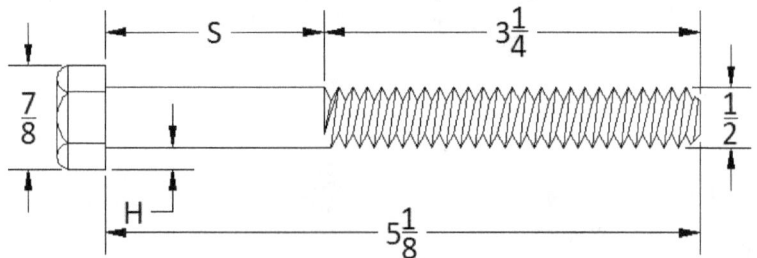

4. If the drawing of the part below is scaled up by a multiple of five, find the new dimensions. Measurements are in inches. Note: An R in drafting denotes the radius of a circle.

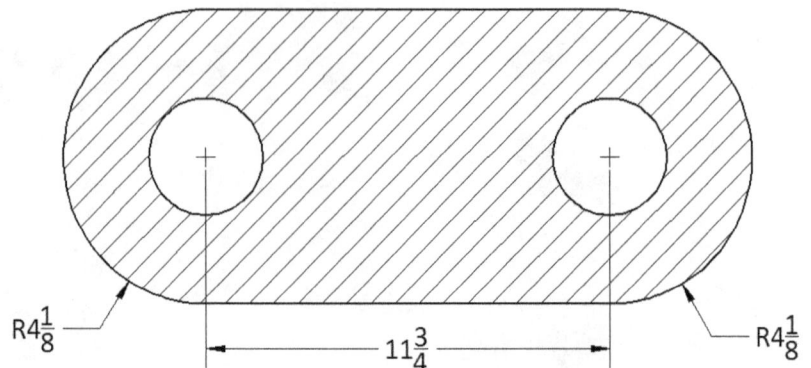

5. If the drawing of the link below is scaled up by a factor of seven, find the new dimensions for the <u>overall</u> width and height. Measurements are in inches. Note: An R in drafting denotes the radius of a circle.

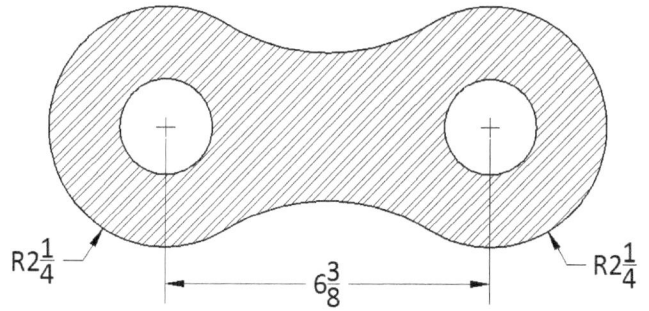

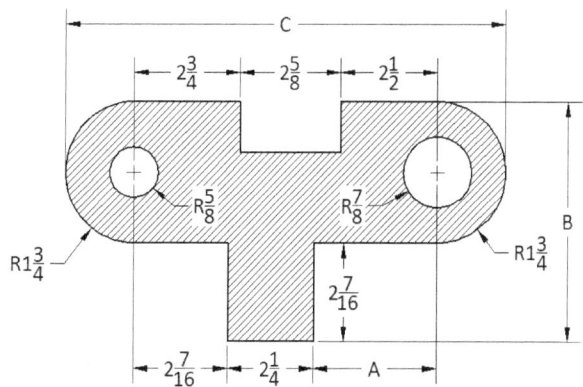

$R2\frac{1}{4}$ $6\frac{3}{8}$ $R2\frac{1}{4}$

6. a) A fence is to have seven boards between two posts so that the space <u>between</u> each board is the same. Calculate the distance between each board. Answer as a fraction.

b) If the carpenter decided the distance between boards was too large, would it be possible to add an extra board by decreasing the space between each board? Explain. Note: dashes are commonly placed between whole numbers and fractions so $21\frac{7"}{8}$ is not mistaken for $\frac{217"}{8}$.

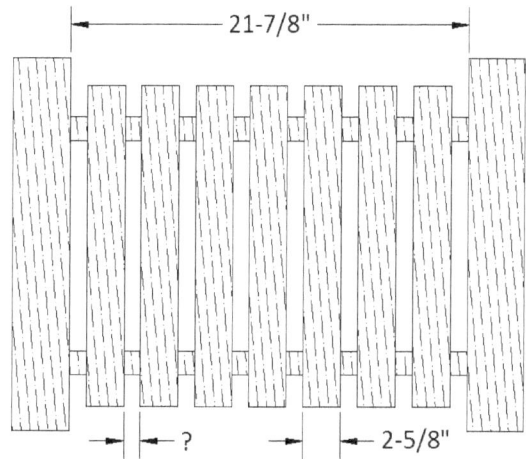

21-7/8"

?   2-5/8"

7. Calculate the missing dimensions A, B, and C. Measurements are in inches. Note: An R in drafting denotes the radius of a circle.

C $2\frac{3}{4}$ $2\frac{5}{8}$ $2\frac{1}{2}$

$R\frac{5}{8}$ $R\frac{7}{8}$

$R1\frac{3}{4}$ $R1\frac{3}{4}$

B

$2\frac{7}{16}$

$2\frac{7}{16}$ $2\frac{1}{4}$ A

8. A stair stringer is cut with a $7\frac{9"}{16}$ rise and a $11\frac{5"}{8}$ run. Code stipulates that all rises must be the same height and all runs must be the same length to avoid being a trip hazard. Calculate the total rise and total run for the stringer with nine steps.

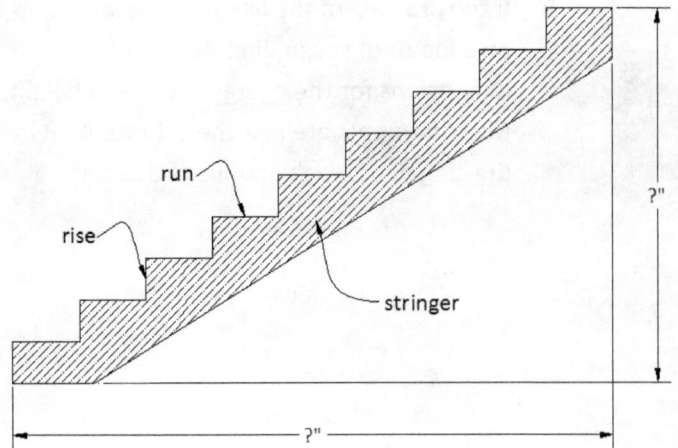

9. A manufacturer wants to layout the parts on a sheet of steel in 14 rows and 6 columns with a ½" space between each row and column. Calculate the width (A) and length (B) of the sheet that will be needed to produce the parts. How many parts will this design produce? Measurements are in inches.

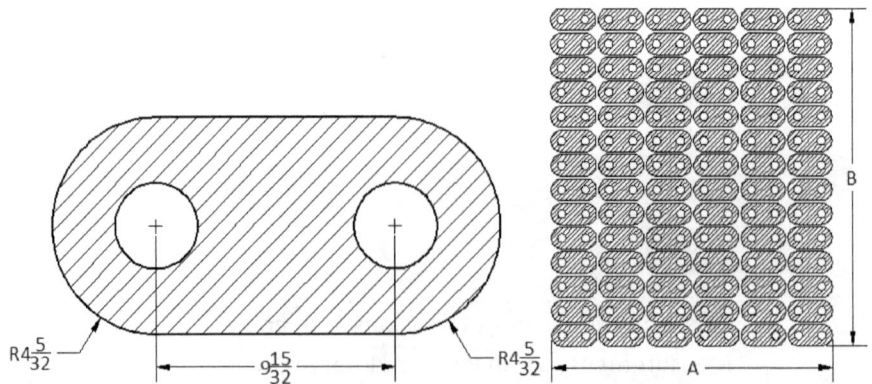

10. A manufacturer wants to lay out the parts on a sheet of plastic in 12 rows and 8 columns with a .15-inch space between each row and column. Calculate the width (A) and length (B) of the sheet of that will be needed to produce the parts. How many parts will this design produce? Measurements are inches.

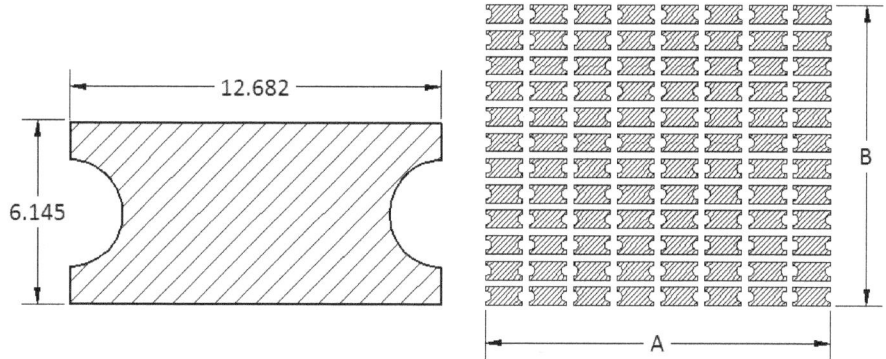

11. Calculate the area and perimeter of a rectangular sheet of glass that is $12\frac{3}{8}"$ x $7\frac{1}{4}"$, answer as a fraction.

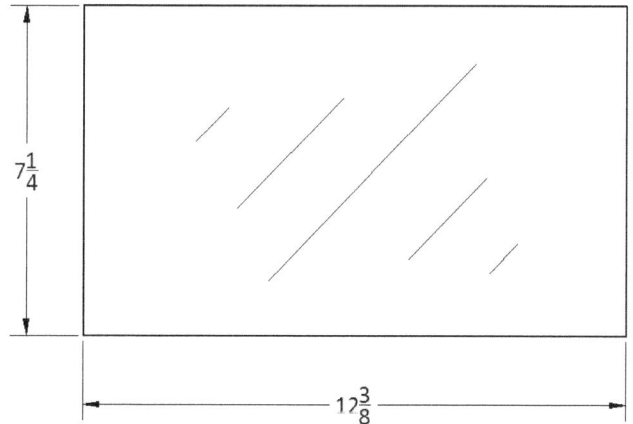

12. If the drawing of the link below is scaled down by a factor of four (divide by four), find the new dimensions for the <u>overall</u> width and height. Measurements are in inches.

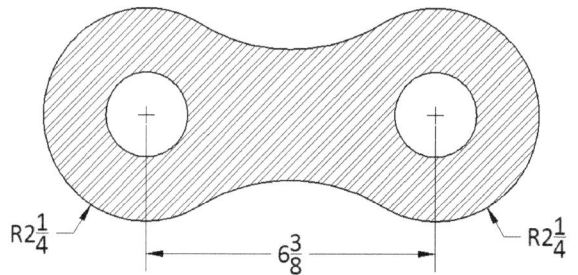

15. In electronics large or small quantities are written using metric prefixes. For example:

$3.2 \ \mu A = 3.2 \times 10^{-6} \ A = 3.2 \times .000001 \ A = .0000032 \ A$

$8.3 \ k\Omega = 8.3 \times 10^{3} \ \Omega = 8.3 \times 1000 \ \Omega = 8300 \ \Omega$

Notice that the power of 10 moves the decimal left or right according to the sign of the number.

Expand the following without prefixes:

Common Electronics Units	
Unit	Symbol
W	Watt
Ω	Ohm
V	Volt
A	Amp
F	Farad
H	Henry
Hz	Hertz

a) 6.32 GV

b) 9.8 nA

c) .47 μF

d) 14.2 MW

Table of Common Metric Prefixes			
Metric Prefix	Symbol	Power of 10	Factor
Tera	T	10^{12}	1,000,000,000,000
Giga	G	10^{9}	1,000,000,000
Mega	M	10^{6}	1,000,000
kilo	k	10^{3}	1,000
milli	m	10^{-3}	.001
micro	μ	10^{-6}	.000001
nano	n	10^{-9}	.000000001
pico	p	10^{-12}	.000000000001

16. Scientific Notation is a number between 1 and 10 times a power of ten.

Engineering Notation is a modification of Scientific Notation where the number is between 1 and 1000 and the power of 10 is always a multiple of three, making it easy to write the number using a metric prefix. Consider the two examples in the table and fill in the rest using the table of common metric prefixes.

Standard Form	Scientific Notation	Engineering Notation	Metric Prefix
12,300 Ω	1.23×10^{4}	12.3×10^{3}	12.3 kΩ
.000072 A	7.2×10^{-5}	72×10^{-6}	72 μA
480,000 Hz			
.00043 V			
7,500 W			
.000000042 F			
62,000,000 H			

Table of Common Metric Prefixes			
Metric Prefix	Symbol	Power of 10	Factor
Tera	T	10^{12}	1,000,000,000,000
Giga	G	10^{9}	1,000,000,000
Mega	M	10^{6}	1,000,000
kilo	k	10^{3}	1,000
milli	m	10^{-3}	.001
micro	μ	10^{-6}	.000001
nano	n	10^{-9}	.000000001
pico	p	10^{-12}	.000000000001

17. Consider a rod being cut to a length. The length can never be perfect and so must be cut for accuracy based on a desired tolerance. Tolerance is the amount that a measurement can vary above or below a design size.

Design Length	Tolerance	Maximum Length	Minimum Length
17.317 in	$\pm$ .034 in		
18.12 cm	$\pm$ .15 cm		
$26\frac{5}{8}$ in	$\pm \frac{1}{32}$ in		
3.2 meters	$\pm$ .06 cm	cm	cm

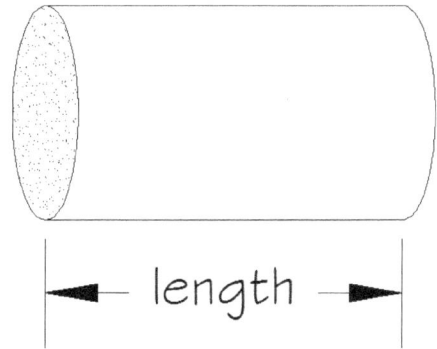

length

18. Calculate the difference between the design size and the actual size in the chart. Actual sizes that are less than the design size should be denoted with negative numbers.

Design Size	Actual Size	Difference
$14\frac{3}{16}$ in	$13\frac{29}{32}$ in	
18.43 mm	18. 38 mm	
17.972 in	18.013 in	

19. **Challenge Problem:** A manufacturer wants to lay out parts on a sheet of steel that is 120" x 68". If the part is 2.34" wide and 1.87" tall, calculate the number of rows and columns that will fit if there is a .6" space between each part. How many parts will this design produce? Note: the drawing is not to scale.

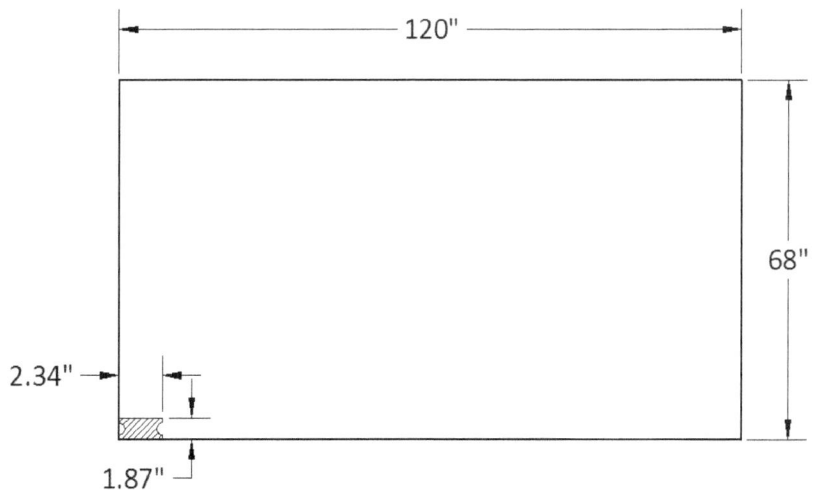

120"

68"

2.34"

1.87"

20. **Challenge Problem:** A manufacturer wants to lay out parts on a sheet of steel that is 108" x 60". If the part is $2\frac{5"}{16}$ wide and $1\frac{1"}{8}$ tall, calculate the number of rows and columns that will fit if there is a $\frac{3"}{4}$ space between each part. How many parts will this design produce? Note: the drawing is not to scale.

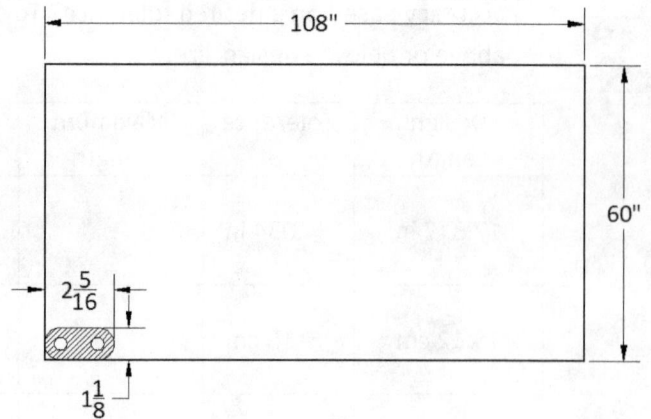

21. Consider the diagram. Convert the given dimensions to fractions to the nearest $\frac{1}{100}$ inch (do not simplify). Using a tolerance of $\pm\frac{2}{100}$ inch, determine the maximum and minimum allowable dimensions.

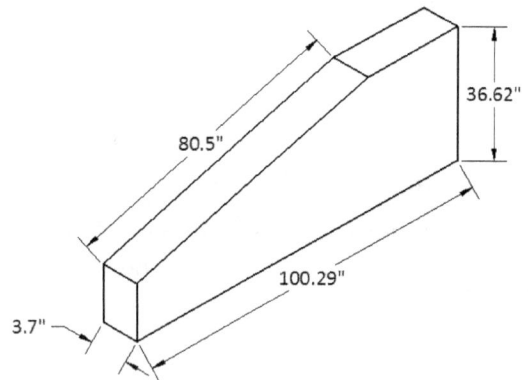

Design Length	Tolerance	Fraction	Maximum Length	Minimum Length
3.7 in	$\pm\frac{2}{100}$ in			
36.62 in	$\pm\frac{2}{100}$ in			
80.5 in	$\pm\frac{2}{100}$ in			
100.29 in	$\pm\frac{2}{100}$ in			

22. Consider the diagram. Determine the maximum and minimum allowable dimensions for the given tolerances in the table below. Answers should be simplified. Note: the symbol Ø indicates the diameter of the circle.

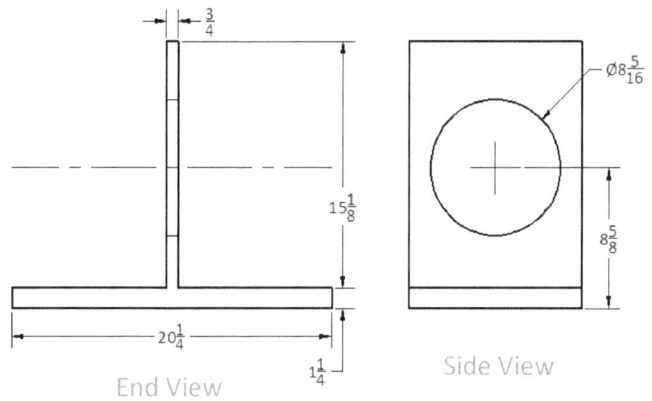

End View

Side View

Isometric View

Design Dimension	Tolerance	Maximum Dimension	Minimum Dimension
$\frac{3}{4}$ in	$\pm\frac{1}{8}$ in		
$1\frac{1}{4}$ in	$\pm\frac{1}{8}$ in		
$8\frac{5}{8}$ in	$\pm\frac{1}{8}$ in		
$8\frac{5}{16}$ in	$\pm\frac{1}{16}$ in		
$15\frac{1}{8}$ in	$\pm\frac{1}{8}$ in		
$20\frac{1}{4}$ in	$\pm\frac{1}{8}$ in		

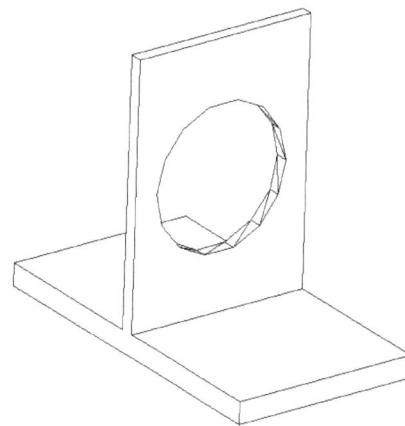

1.2: Measurement

Simply put, measurement is the language of industry. A familiarity with the metric and standard systems of measurement is essential in creating and reading blueprints.

The standard system is cumbersome, based on the size of thumbs, arms and feet and divided and named without any thought as to the trouble it would cause us here in the 21st century. There are 3 feet in a yard, 12 inches in a foot, and 5280 feet in a mile; and this without even mentioning rods, chains, furlongs or fathoms. The key for most applications is an understanding of feet, inches and parts of an inch. For most applications, measuring accurately to the nearest 32nd of an inch is more than sufficient. Most rulers and tape measures divide the inch into 16 equal parts so it takes practice to measure to the nearest 32nd. The number of divisions also evolved slowly as the need for accuracy increased with technology. First the inch was divided in half (1/2), then the halves in half for quarters (1/4), then the quarters in half for eighths (1/8), next the eighths in half for sixteenths (1/16), and finally the sixteenths in half for thirty-seconds (1/32).

In the ruler below, notice the inches are divided into 16 parts, so each division represents $\frac{1"}{16}$:

Example 1.2.1: Naming measures on a standard ruler

Name the measurements:

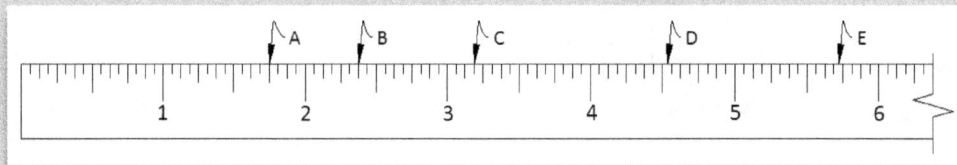

Final Answers: $A = 1 + \frac{12}{16}$ or $1\frac{3"}{4}$

$B = 2 + \frac{6}{16}$ or $2\frac{3"}{8}$

$C = 3 + \frac{3}{16}$ or $3\frac{3"}{16}$

$D = 4 + \frac{1}{2} + \frac{1}{32}$ or $4 + \frac{16}{32} + \frac{1}{32}$ or $4\frac{17"}{32}$

$E = 5 + \frac{3}{4} - \frac{1}{32}$ or $5 + \frac{24}{32} - \frac{1}{32}$ or $5\frac{23"}{32}$

Note: the marks indicating half inches are shorter than those indicating inches. The marks for quarters are shorter than those for halves; marks for eighths are even shorter, and so on. With practice you should be able to identify a measurement like $3\frac{5"}{8}$ without counting individual lines.

A notable exception to this degree of accuracy in the standard system exists in manufacturing where the inch is divided into 1000 equal parts. A measurement off by $\frac{1"}{32}$ is beyond tolerance for pistons in an automobile engine, for example. One nice advantage, beyond the increased level of accuracy, is that a measurement of $5\frac{243"}{1000}$ can easily be written in decimal form as 5.243". Admittedly this may disappoint manufacturing students not to be able to work with fractions, but some sacrifices must be made in the name of progress!

It is interesting to note that 1000 marks between each inch would be impossible to distinguish on a typical ruler, so a brilliant device called a vernier caliper is used. Half an hour with someone familiar with this device will enable you to confidently measure with it.

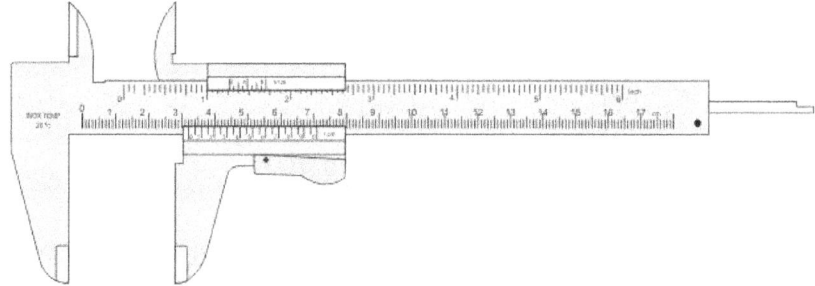

As you are probably aware, the entire world, save the Englishman, recognized the ease of dividing measurements by powers of ten (tenths, hundredths, thousandths, etc.). The basic unit of length in the metric system is the meter, and all other units are found by multiplying or dividing the meter by a power of 10. Measurements in the metric system can always be written in decimal form, making calculations much easier. The same prefixes are used in the metric system, regardless of whether one is measuring a length, weight, or liquid measure.

Common Metric Prefixes

Prefix	Meaning
Kilo-	1 thousand = 1000
Centi-	1 hundredth = $\frac{1}{100}$
Milli-	1 thousandth = $\frac{1}{1000}$
Micro-	1 millionth = $\frac{1}{1000000}$

7.8 kilometers is 7800 meters and

346 centimeters is 3.46 meters, for example.

In the typical metric ruler below, the numbers represent centimeters (cm) which are divided into 10 parts. Each division represents a millimeter (mm):

Example 1.2.2: Naming measures on a metric ruler

Name the measurements:

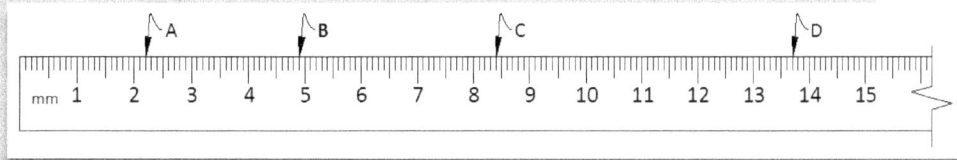

Final Answers: A = 2.2 cm or 22 mm

B = 4.9 cm or 49 mm

C = 8.4 cm or 84 mm

D = 13.7 cm or 137 mm

The ability to convert back and forth from decimals to fractions is essential since many of the calculations that we will explore in this text involve formulas that will result in decimal answers. Our standard system of measurement requires that numbers be in fraction form so that they can be measured. Consider the following practical example:

Example 1.2.3: Applying skills with fractions and decimals

A cabinet is to be built with 3 equal spaces for sinks and drawers. Find the size of each space if the overall size is $112\frac{3}{4}$ ", rounded to the nearest 32nd of an inch.

Solution:

It is simple to use a calculator to convert a fraction into a decimal. A fraction bar is a symbol for division so $\frac{3}{4}$ = 3 ÷ 4 = .75. This should come as little surprise since with money three quarters is 75 cents. With a calculator then 112.75 ÷ 3 ≈ 37.5833, rounded to four decimal places. It is a little more than 37", but how many 32^{nds} of an inch is .5833? Restated as a simple equation it becomes apparent, $\frac{?}{32}$ = .5833. Multiplying .5833 x 32 and rounding gives the answer 18.7 which is close to 19 so it is nearest to $\frac{19}{32}$.

Final Answers: $37\frac{19}{32}$ "

One more example just to be sure you have it:

Example 1.2.4: Skills with fractions and decimals

A $91\frac{3}{8}$ "piece of steel is to be divided into 5 equal pieces. Find the size of each piece rounded to the nearest 16th of an inch.

Solution:

Again, $\frac{3}{8}$ = 3 ÷ 8 = .375. With a calculator then 91.375 ÷ 5 = 18.275. Notice that in the previous example the decimal required rounding, since the decimal never stops, whereas this time it does not. Mathematicians get pretty excited about this difference since you cannot technically get the previous example correct in decimal form. Again the simple equation, $\frac{?}{16}$ = .275. Multiplying .275 by 16 and rounding gives the answer. .275 x 16 = 4.4 which is closer to 4 so it is nearest to $\frac{4}{16}$ or $\frac{1}{4}$.

Final Answers: $18\frac{1}{4}$ "

Section 1.2: Measurement

1. Find the measurements indicated by the arrows on the standard ruler.

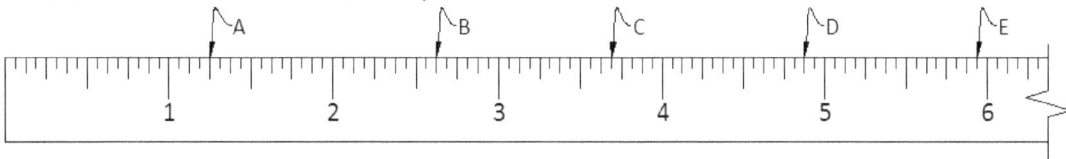

2. Find the measurements indicated by the arrows on the standard ruler. All measurements fall between 16^{ths}, answer to the nearest 32nd of an inch.

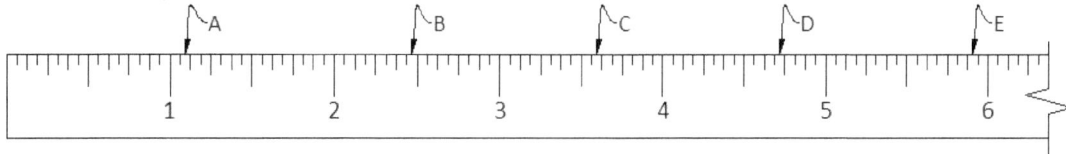

3. Find the measurements indicated by the arrows on the metric ruler in centimeters.

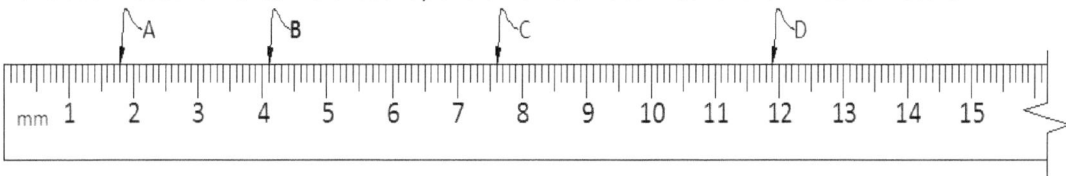

4. Find the measurements indicated by the arrows on the metric ruler in millimeters.

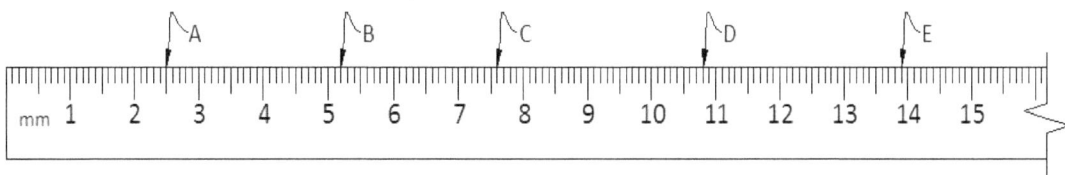

5. Measure the dimensions A through S on the part to the nearest millimeter.

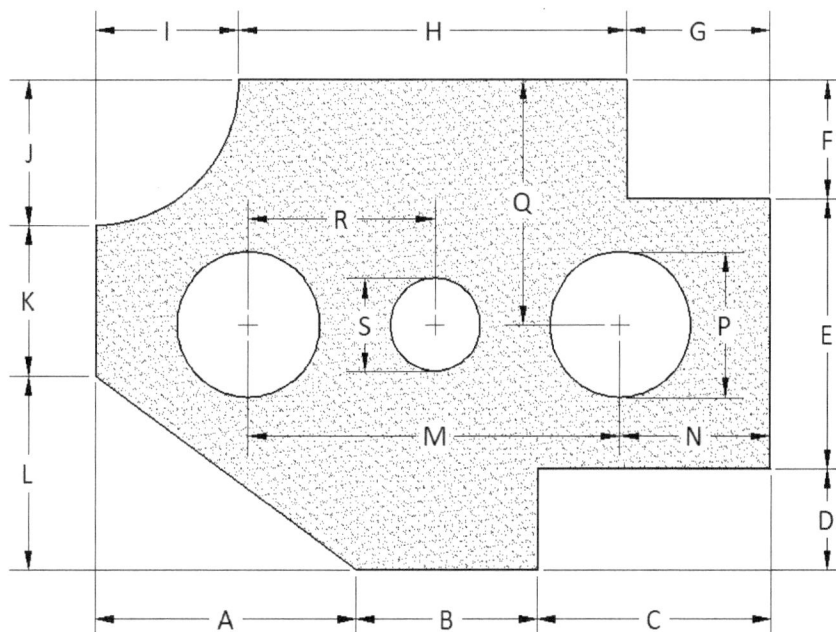

6. Measure the dimensions A through S on the part to the nearest 32nd of an inch. Answer as a reduced fraction.

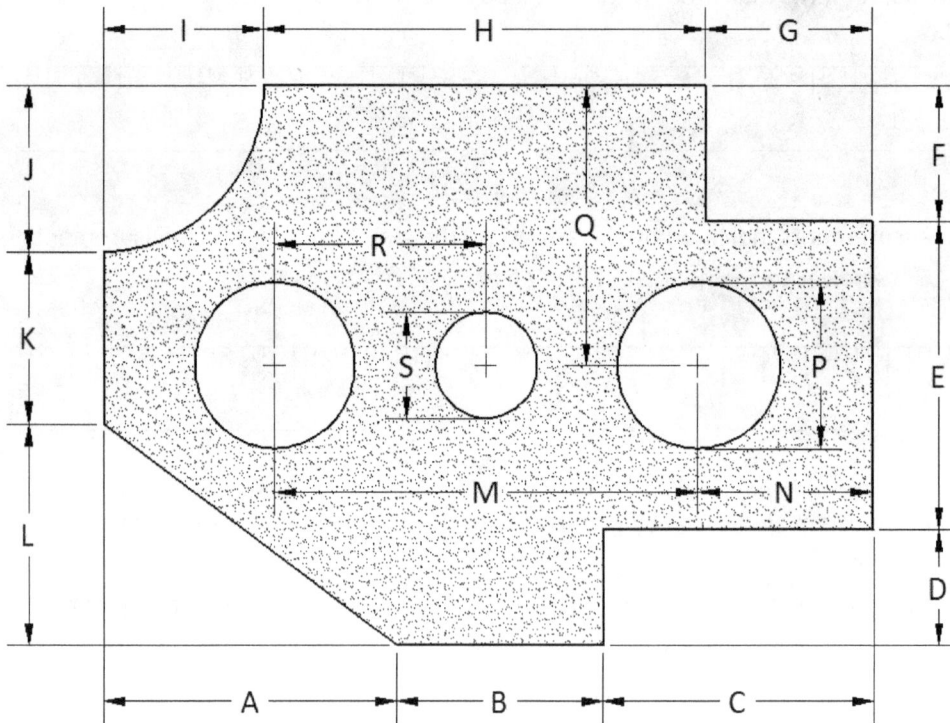

7. A fence is to have 11 boards between two posts so that the space <u>between</u> each board is the same.   Calculate the distance between each board rounded to the nearest 16th of an inch.

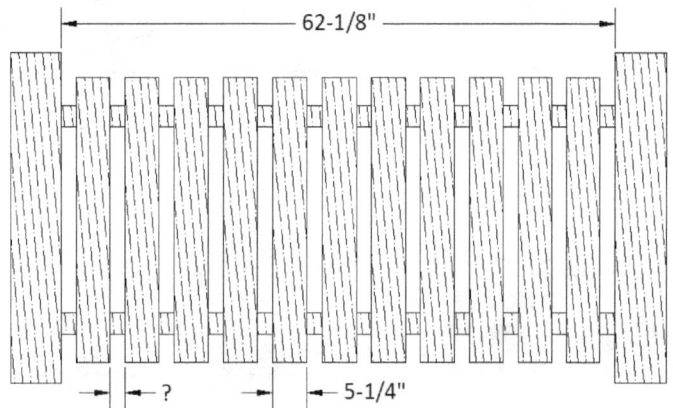

62-1/8"

? 5-1/4"

8. A fence is to have $3\frac{1}{2}"$ wide boards between two posts that are $94\frac{1}{4}"$ apart so that the space <u>between</u> each board is the same. Calculate the number of boards and the spacing so that the spacing between the boards is as small as possible. Round your answer for the spacing to the nearest 16th of an inch.

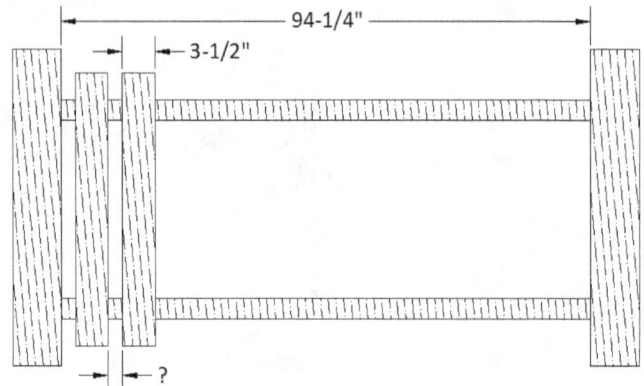

94-1/4"

3-1/2"

?

9. Copper pipe is manufactured in three types according to wall thickness. Use the outside and inside diameter measurements from the chart to calculate the wall thickness as a decimal rounded to three places <u>and</u> as a fraction rounded to the nearest 32nd of an inch. All measurements are in inches.

    a)   Size 3/8   Type M

    b)   Size 5/8   Type L

    c)   Size $1\frac{1}{4}$ Type K

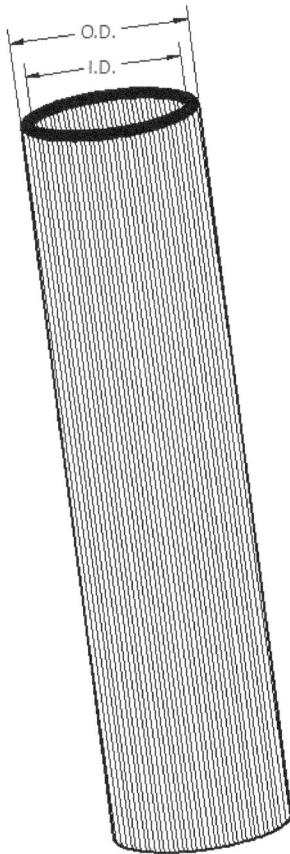

Size	outside diameter (O.D.)	Type		
		K	L	M
		inside diameter (I.D.)		
3/8	1/2	0.402	0.430	0.450
1/2	5/8	0.528	0.545	0.569
5/8	3/4	0.652	0.668	0.690
3/4	7/8	0.745	0.785	0.811
1	1-1/8	0.995	1.025	1.055
1 ¼	1-3/8	1.245	1.265	1.291
1 ½	1-5/8	1.481	1.505	1.527
2	2-1/8	1.959	1.985	2.009

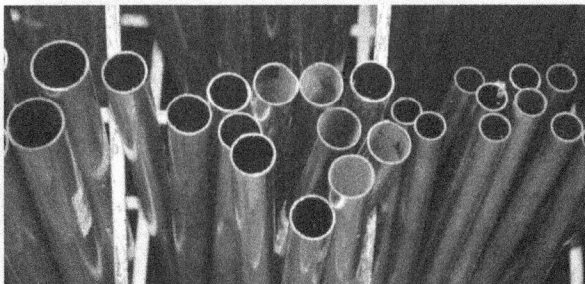

10. Fill in the missing columns in the pilot-hole size chart below. Loose pilot holes should be fractions rounded up to the nearest 32nd of an inch so that the screw can slide through. Tight pilot holes should be rounded down to the nearest 32nd of an inch so the screw threads have wood to grip. Answer as a reduced fraction.

Wood Screw Pilot Hole Sizes			
Screw Size	Thread Diameter	Loose Pilot Hole	Tight Pilot Hole
4	.112		
6	.138		
11	.203		
14	.242		

11. If the first aid sign below is scaled down by a factor of nine, find the new dimensions for the <u>overall</u> width and height rounded to the nearest 16th of an inch. Assume the shape to be symmetrical and all dimensions to be in inches.

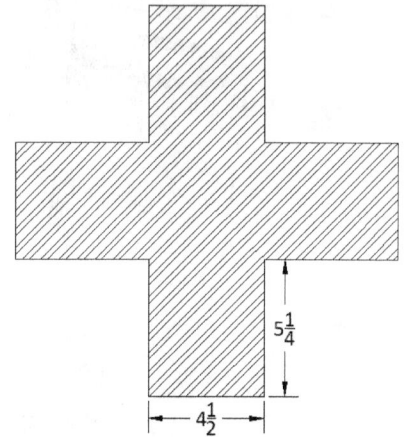

$5\frac{1}{4}$

$4\frac{1}{2}$

12. A stair stringer is to be cut with nine rises to reach a deck that is 68 inches above the ground. Construction code stipulates that each rise must be equal so that the stair case will not be a trip hazard. Calculate the height of each rise rounded to the nearest 16th of an inch.

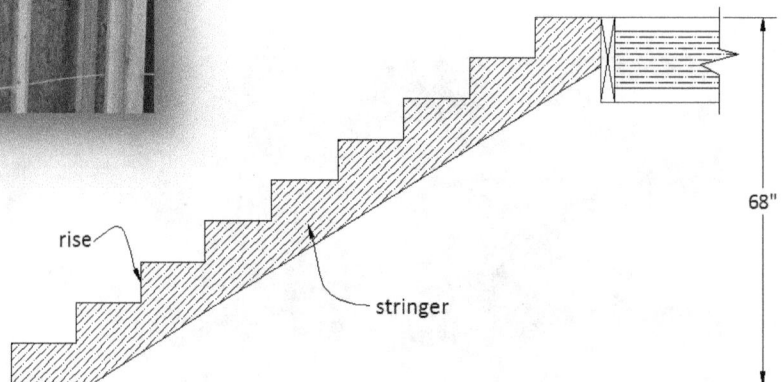

rise

stringer

68"

13. A stair stringer is to be cut to reach a 2nd story floor that is 106 inches above the ground floor. Construction code stipulates that each rise must be equal so that the stair case will not be a trip hazard. Calculate the whole number of rises so the height of each rise is as near the ideal of seven inches as possible. Next, use the whole number of rises to calculate the height of each rise rounded to the nearest 16th of an inch.

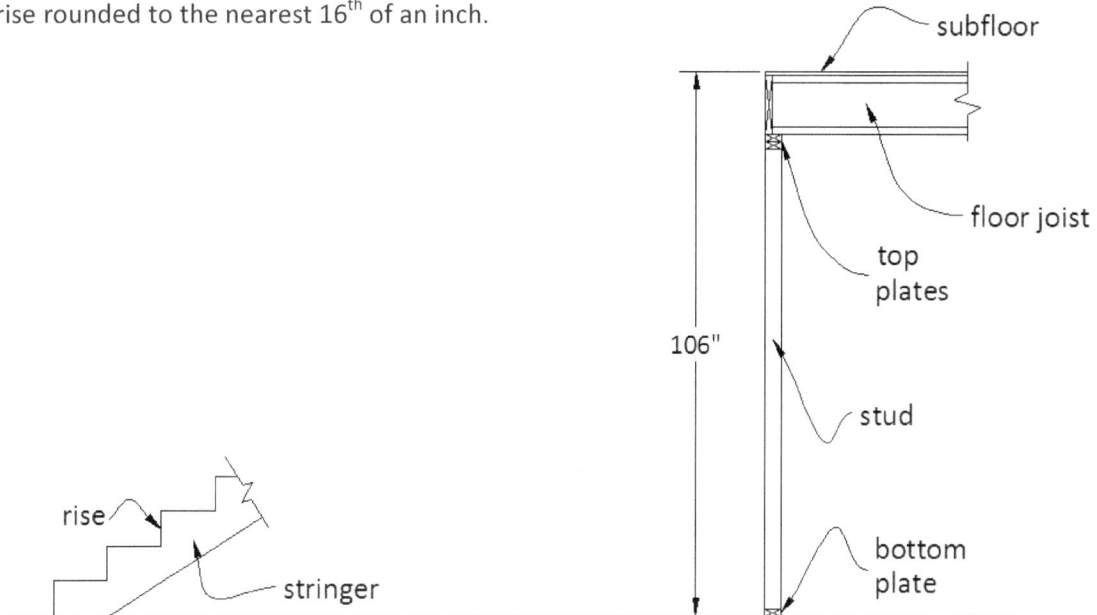

14. A stair stringer is to be cut to reach a 2nd story floor in a house with nine foot ceilings. The distance from the 1st to the 2nd floor can be calculated by adding up the height of each material. Bottom plate $1\frac{1}{2}''$, stud $104\frac{1}{4}''$, two top plates $1\frac{1}{2}''$ each, floor joist $11\frac{7}{8}''$, and subfloor $\frac{3}{4}''$. Construction code stipulates that each rise cannot exceed eight inches and must be equal so that the stair case will not be a trip hazard. Calculate the number of rises and the height of each rise (rounded to the nearest 16th of an inch) for all possible designs where the rise is between six and eight inches.

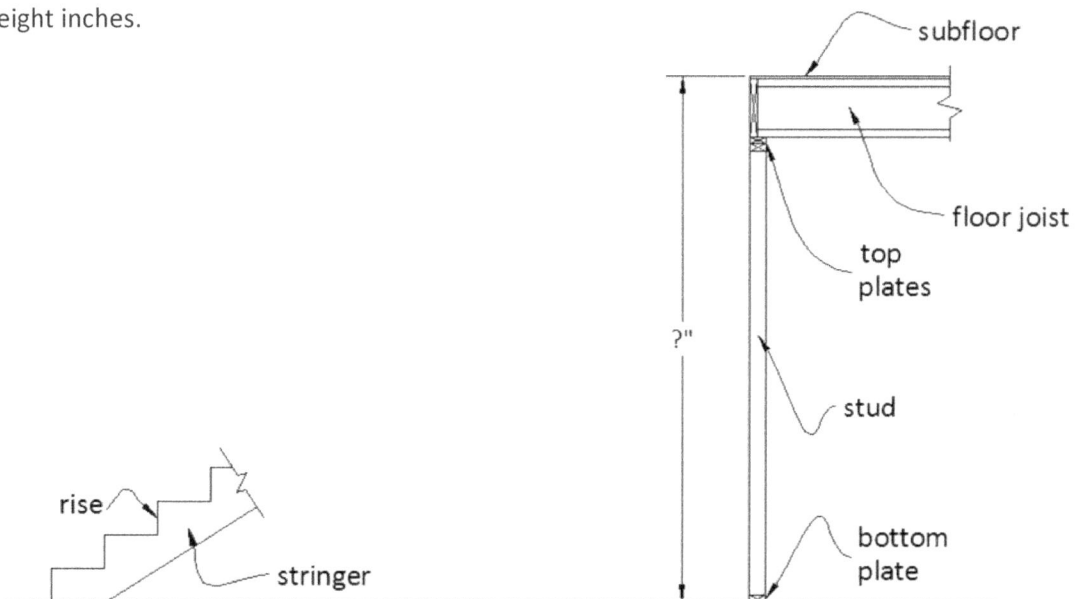

15. A manufacturer must describe a dimensioned part according to its x & y coordinates for a CNC (computer numerically controlled) machine to produce it. These coordinates are measured relative to a datum. The x-coordinate is the horizontal distance from the datum and the y-coordinate is the vertical distance from the datum. Fill in the chart that locates the holes and points that define the part.

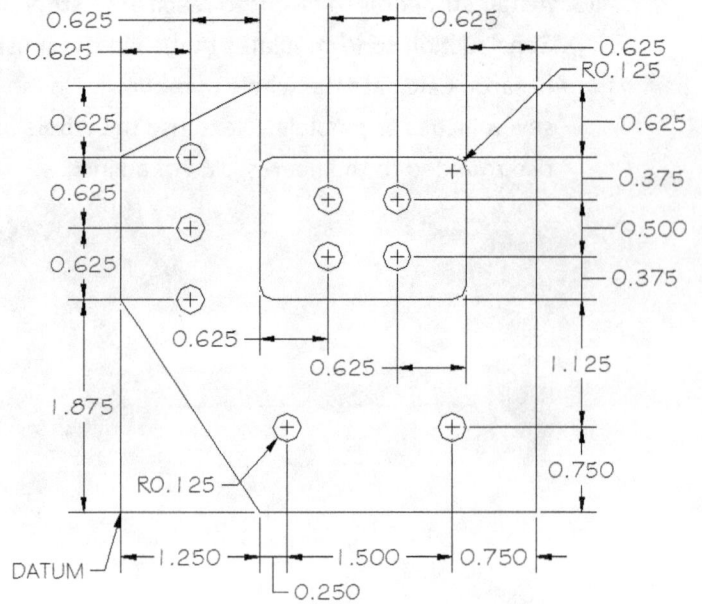

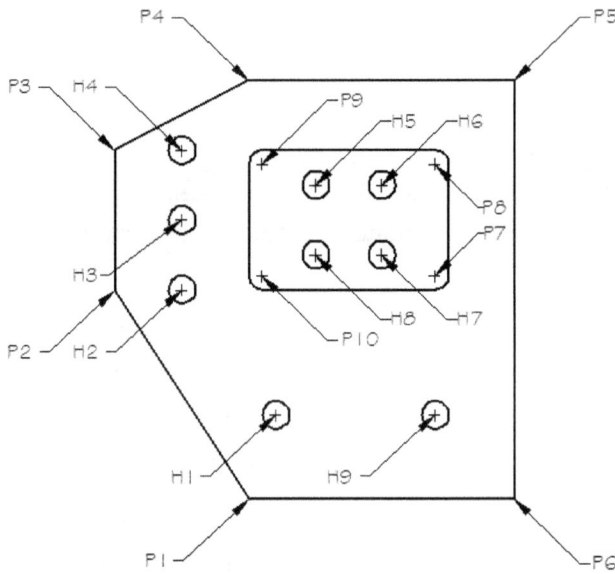

	X - Coordinate	Y - Coordinate
P2		
P4		
P6		
P9		
H3		
H6		
H8		
H9		

16. A manufacturer must describe a dimensioned part according to its x & y coordinates for a CNC (computer numerically controlled) machine to produce it. These coordinates are measured relative to a datum. The x-coordinate is the horizontal distance from the datum, right is positive and left is negative. The y-coordinate is the vertical distance from the datum, up is positive and down is negative. Fill in the chart that locates the holes and points that define the part.

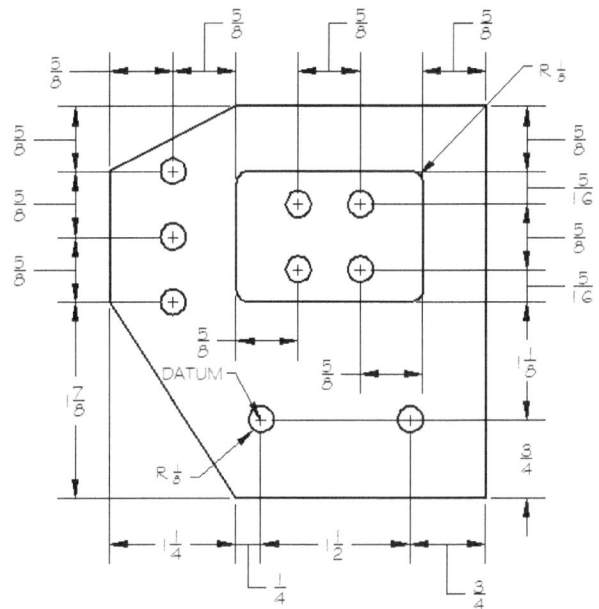

	X - Coordinate	Y - Coordinate
P1		
P3		
P5		
P7		
P8		
P10		
H1		
H2		
H4		
H5		
H7		

17. Four pieces of equal length are cut from the 20 inch length of round stock. If the saw takes 3/16", calculate the length of the left over piece as a fraction. All dimensions are in inches.

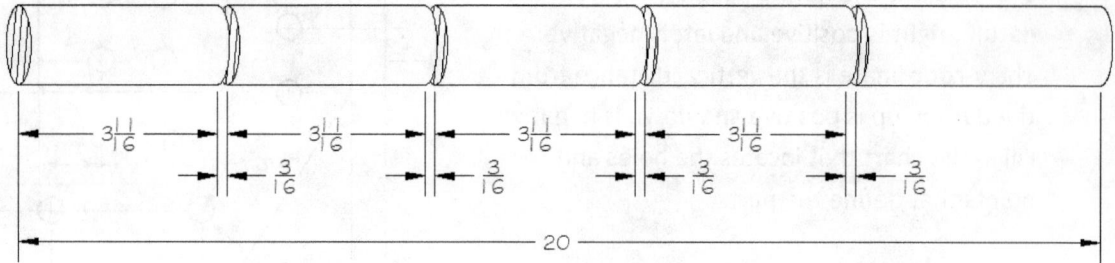

$3\frac{11}{16}$ $3\frac{11}{16}$ $3\frac{11}{16}$ $3\frac{11}{16}$

$\frac{3}{16}$ $\frac{3}{16}$ $\frac{3}{16}$ $\frac{3}{16}$

20

18. A length of round stock is to be cut into six equal lengths using a saw that removes 3/8" with each cut. Calculate the length of the equal piece. All dimensions are in inches.

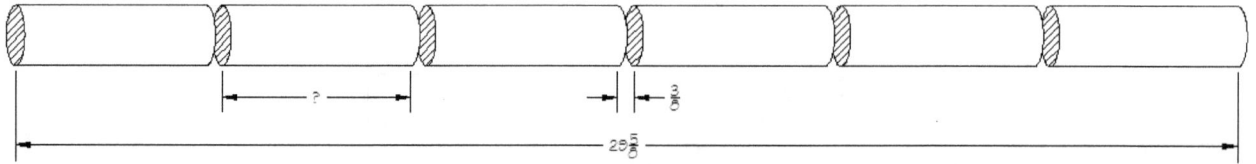

? $\frac{3}{8}$

$29\frac{5}{8}$

19. A section of a gear is shown. Given the formulas below and the circular pitch in the chart, determine the working depth, clearance, and tooth thickness. Answer as decimals rounded to 3 decimal places.

Working Depth = .6366 x Circular Pitch
Clearance = .05 x Circular Pitch
Tooth Thickness = .5 x Circular Pitch

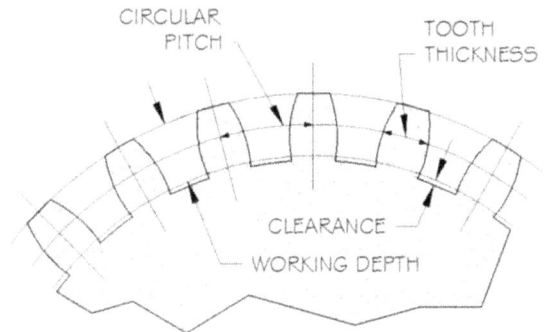

Circular Pitch	Working Depth	Clearance	Tooth Thickness
.398 inches			
2.65 cm			
$\frac{9}{16}$ inches			
$\frac{472}{1000}$ of an inch			

1.3: Ratio, Proportion and Percent

Proportions are one of the simplest and most powerful tools that math has to offer. I recently replaced the ridge cap on a colleague's roof. The lineal footage to be replaced measured 256 feet. How many boxes to order? My problem was easily solved making a proportion by setting 2 ratios equal to each other. Each box of ridge cap contained 20 lineal feet so: $\frac{1\ box}{20\ ft} = \frac{?\ boxes}{256\ feet}$. The answer was 12.8 boxes so I ordered 13 boxes. Each box of ridge cap contained 30 individual ridge caps which are attached with 2 nails each. I would need 60 nails per box times 13 boxes, so 390 nails. Counting this many nails would take an unreasonable amount of time so the salesman used a proportion based on the table to sell me the nails by weight. I used $1\frac{1"}{2}$ nails so:

$\frac{1\ lb}{180\ nails} = \frac{?\ lbs}{390\ nails}$. He sold me 2.2 lbs.

Length	Gauge	Head	Nails/Lb.
7/8"	11	3/8"	272
1"	11	3/8"	250
1-1/4"	11	3/8"	202
1-1/2"	11	3/8"	180
1-3/4"	11	3/8"	156
2"	11	3/8"	136

The power of the proportion lies in establishing one ratio relating two quantities you are interested in, then equating that to another. I found that I was able to install 24 feet of ridge cap in 32 minutes. I wanted to know if I would be done in time to pick up my daughter from school. A proportion allowed me to predict how long it would take to finish the job.

Example 1.3.1: Roofing application

How much time will it take to install 256 feet of ridge cap if the first 24 feet are installed in 32 minutes?

Solution:

Set up a proportion: $\frac{32\ min}{24\ ft} = \frac{?\ min}{256\ ft}$.

$\frac{32}{24} = \frac{t}{256}$ rewrite without the units

$32 \cdot 256 = 24 \cdot t$ We will discuss solving equations more carefully in chapter 2, for now, cross multiplying eliminates the fractions. The dot (•) is a common symbol for multiplication in algebra since the usual symbol (x) is also used as a letter.

$8192 = 24t$ simplify

$t \approx 341$ divide both sides by 24

Final Answer: total time for the job 341 minutes. I had already worked for 18 minutes so that left about 323 minutes.

Side note 1: I didn't factor in that I am 44 years old I cannot bend over for $5\frac{1}{2}$ hours straight. I was late.

Side note 2: I only estimated this since the situation did not call for a great deal of accuracy, and more importantly, carrying your calculator with you in life is universally regarded as nerdy.

Consider another example:

Example 1.3.2: Sloping concrete for drainage

Concrete contractors typically slope garage floors at ¼" per foot so that water will drain off. If the floor is 26 feet long, find the total amount of fall or drop in the floor.

Solution:

Set up a proportion: $\frac{\frac{1}{4} \text{ in}}{1 \text{ ft}} = \frac{? \text{ in}}{26 \text{ ft}}$.

$\frac{\frac{1}{4}}{1} = \frac{x}{26}$ without the units

$26 \bullet \frac{1}{4} = 1 \bullet x$ cross multiply

$6\frac{1}{2} = x$ simplify

Final Answer: The total fall in the floor will be $6\frac{1"}{2}$.

Another excellent use for proportions arises when converting decimal calculations to fractions in the standard system of measurement. This was briefly considered in section 1.2.

Example 1.3.3: Roof slope application

A roof slopes with a rise of 7 and run of 12. Find the rise accurate to the nearest 16th of an inch for a run of $118\frac{5"}{8}$.

Solution:

$\frac{7}{12} = \frac{x}{118.625}$ both rises on the top of the fraction, both runs on the bottom

$7 \bullet 118.625 = 12 \bullet x$ cross multiply

$830.375 = 12x$ simplify

$x \approx 69.198$ divide both sides by 12

Although this answer is correct, it is not measurable with a typical ruler.

We need to figure out how many 16ths are in .198, which can also be solved with a proportion.

$\frac{.198}{1} = \frac{?}{16}$ how many 16ths equal .198?

$? \approx 3.2$ cross multiply (.198" is approximately $\frac{3"}{16}$)

Final Answer: The roof will rise approximately $69\frac{3"}{16}$.

Another application is found in reading plans:

Example 1.3.4: Scale drawing application

A manufacturer is drawing a plan scaled at $\frac{1"}{8} = 1"$ (meaning the plan is drawn 1/8 the size of the real part). If the part measures 35.348 inches, find the length of the measure on the plan to the nearest 10th of an inch.

Solution:

$\frac{\frac{1}{8}}{1} = \frac{x}{35.348}$ set up a proportion

$x \approx 4.4"$ cross multiply

Final Answer: Draw the part approximately 4.4".

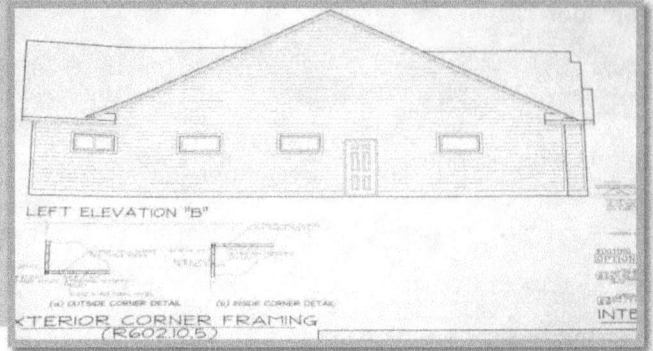

A fraction can have any number in the denominator. A percent is simply a fraction with a denominator of 100. Percents are convenient for comparison because as the name (per-cent) implies, they are always per 100 or $\frac{?}{100}$. The symbol "%" is used in place of the fraction for convenience. If you think about the word percent it should be evident that 32% $= \frac{32}{100}$ = .32. When you see "37%", think "37 hundredths". When you encounter a fraction like $\frac{2}{5}$, keep in mind that changing it to percent means the same thing as changing it to hundredths. Thus, $\frac{2}{5} = \frac{40}{100}$ = 40%.

In example 1.3.1, I noted that I had completed $\frac{24 \text{ ft}}{256 \text{ ft}}$ of my roofing job. If this fraction is changed to a percent, it is easier to have a feel for how much of the job I have completed. One method for changing to a percent is to use a proportion. We must change $\frac{24}{256}$ to hundredths, so we write $\frac{24}{256} = \frac{x}{100}$. Cross-multiplying, we see that 2400 = 256x, so x ≈ 9. From this, we can conclude that $\frac{24}{256}$ is about 9 hundredths, so I had completed about 9% of the job.

A second method for changing a fraction to a percent involves the fraction bar in $\frac{24}{256}$, indicating that 24 is divided by 256. The fraction $\frac{24}{256}$ = .09375 which rounds to .09 or 9%.

Percents are regularly used in the financial aspects of trade.

Example 1.3.5: Costing out a job

A welder agrees to build a trailer for cost plus 18%. Calculate the total bill for a $1,460 job.

Solution:

18% of 1,460	find the amount to add to the cost
$\frac{18}{100} \bullet 1,460$	18 percent = 18 hundredths and "of" is a word implying multiplication
.18 • 1,460	18 hundredths in decimal form
262.8	simplify
1,460 + 262.8	costs plus the added 18%

Final Answer: He will expect to be paid $1,460 + $262.80 = $1,722.80.

Example 1.3.6: Applying a discount

A lumber bill arrives with a note that you can subtract 7% if you pay on time. Calculate the amount you should pay if the bill is for $17,654.

Solution:

7% of 17,654	find the amount to subtract from the bill
.07 • 17,654	"of" means multiply and 7% as a decimal
1,235.78	simplify
17,654 – 1,235.78	bill amount minus the 7% discount

Final Answer: You would pay $17,654 – $1,235.78 = $16,418.22.

Note: A wonderful shortcut: Saving 7% would mean that you are paying 93%. And .93 x 17,654 = 16,418.22, which is the same result.

Dilution is the amount of an added substance divided by the total volume. Suppose that 15 mL of epinephrine are added to 60 mL of anesthetic. Clearly the solution has a total volume of 75 mL. Medically this is described as a $\frac{15}{75}$ or a $\frac{1}{5}$ dilution of epinephrine in anesthetic.

Do not confuse this with ratio. The ratio of epinephrine to anesthetic is 15:60 or 1:4.

Example 1.3.7: Dilutions

Suppose a nurse needs 112 ounces of a $\frac{1}{8}$ dilution of lidocaine in saline. Find the number of ounces of lidocaine and saline that are required.

Solution:

$\frac{1\ part\ lidocaine}{8\ parts\ total\ volume} = \frac{x\ part\ lidocaine}{112\ parts\ total\ volume}$	set up a proportion
112 = 8x	cross multiply
x = 14	divide both sides by 8

Final Answer: 14 ounces of lidocaine and 112 - 14 = 98 ounces of saline

Section 1.3: Ratio, Proportion and Percent

1. Use a proportion to calculate the height that a
   rafter will reach above the wall that it rests on if
   it is sloped at 8/12.

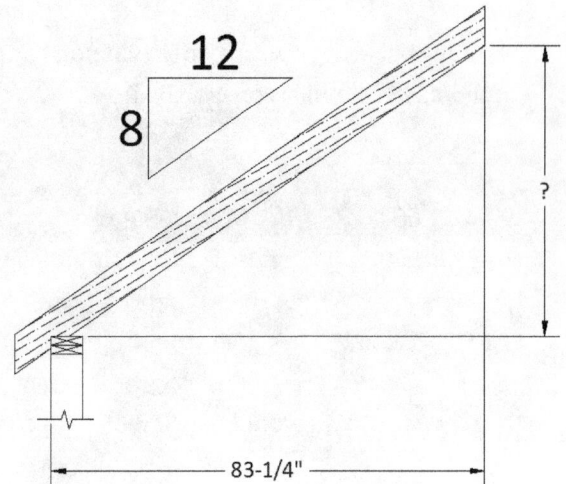

2. 50L of glucose are required for every 1800L of saline. Find the amount of glucose needed for
   4400L of saline. Round to the nearest liter.

3. A slope of ¼" per foot is commonly used by concrete companies to ensure the water doesn't
   pool. Use a proportion to calculate the amount of fall that a garage floor should have if it is 18'-
   6" long. Note: 18'-6" means 18 feet plus 6 inches.

4. Plumbers use a slope of ¼" per foot to ensure the proper flow in waste pipes. Use a proportion
   to calculate the amount of fall that a pipe should have if it is 31'-3" long.

5. Find the time it will take to drain a 1000 mL IV bag if 162 mL drained in 8 minutes. Round to the
   nearest minute.

6. The slope of a hill is 1/8. If an excavator is making a level cut for a house pad that is 54',
   calculate the height of the bank that will result at the uphill side of the cut. Answer in inches.

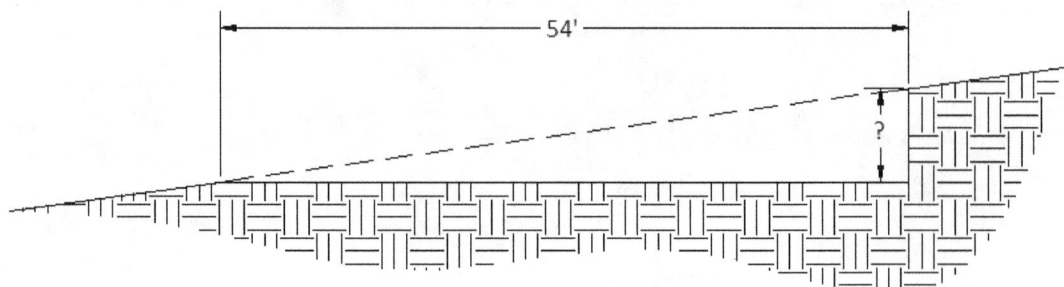

7. In a gable, each trapezoidal piece of lap siding is shorter than the one below by the same amount. A proportion can be used to calculate that amount, allowing a carpenter to cut the pieces without taking measurements. The bottom of each piece of siding is placed 7 inches above the bottom of the piece below. Use the roof's slope of 5/12 and the long point to long point measurement of the first piece, to set up a proportion and calculate the long point to long point measurement of the 2nd piece.

Round your answer to the nearest 16th of an inch.

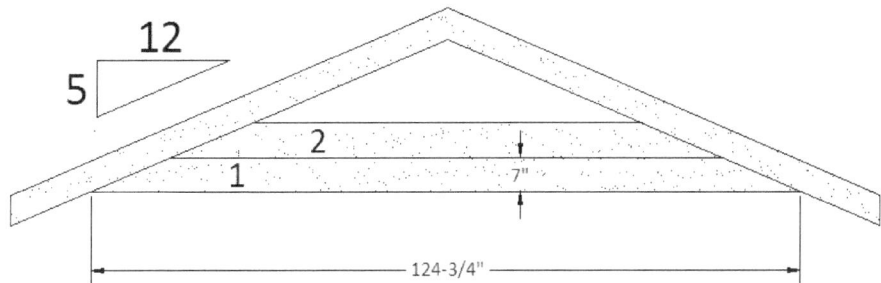

8. A 20 mL portion of a urine sample contains 1.2 mg of a drug. Find the number of milligrams of the drug that would be in the entire 170 mL urine sample.

9. Studs in a framed wall are placed 16 inches apart. A sloped wall presents a challenge in that each stud must be cut to a different length. If the top plate has a slope of 10/12, set up a proportion to calculate the difference in length rounded to the nearest 16th of an inch.

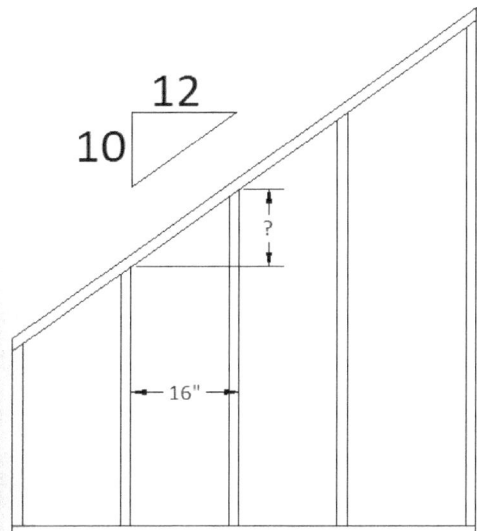

10. The ADA (American Disabilities Act) specifies that a ramp must have a slope of 1/12. If a ramp must attain a height of 14-3/8", calculate the horizontal length of the ramp in inches.

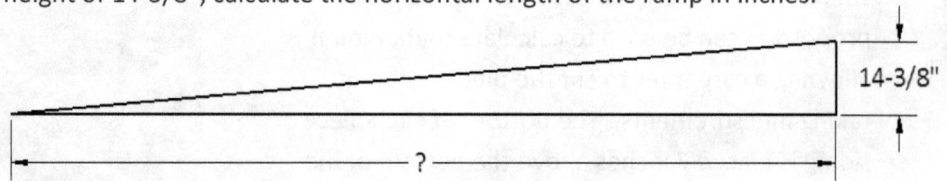

14-3/8"

?

11. A patient adjusted their diet, decreasing their cholesterol count from 198 to 168. Find the percent decrease that this represents.

12. A well produces 18 gallons per minute (GPM). Find the time it will take to fill a 1320 gallon pool rounded to one decimal place.

13. Convert a dose of 132 mL for a 96 pound patient to mL/lb (milliliters per pound), rounded to one decimal place.

14. A car gets 32 miles per gallon (MPG). Find the number of gallons required to travel 857 miles rounded to one decimal place.

15. A car travels at 68 miles per hour (MPH). Find the time necessary to travel 487 miles rounded to one decimal place.

16. Convert a medicine concentration of 8 grams in 22 liters to g/L (grams per liter), rounded to two decimal places.

17. A shaper has a feed rate of 14 feet per minute. Find the time necessary to mill 1700 feet of trim rounded to one decimal place.

18. **Challenge Problem:** Find the amount of each ingredient necessary to make 224 mL of a $\frac{1}{7}$ dilution of glucose in water.

19. A lumber bill for a job totaled $18,400. If a 7% discount was then applied, find the discounted cost for the lumber.

20. If a contractor bills out his work at cost plus 15% (cost + 15% of the cost = total amount), figure the total bill if his costs were $1,340.

21. A 1200 mL IV drip is labeled to contain 8% of a laxative. Find the number of mL of the IV drip that should be administered if a patient is to receive 26 mL of the laxative.

22. If the part below is scaled up 18%, find all three new dimensions.

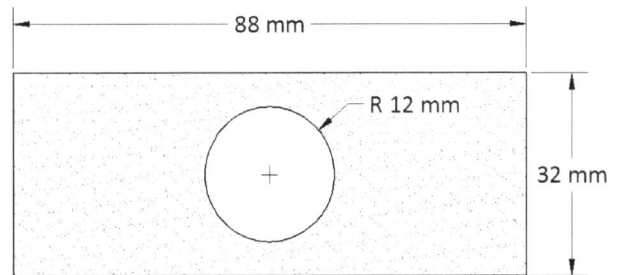

88 mm

R 12 mm

32 mm

23. A patient is administered 320 cm^3 of a saline and water solution that is labeled as 12% saline. Find the amount of each ingredient the patient will receive. Round to the nearest cubic centimeter.

24. If the part below is scaled down 24%, find all three new dimensions.

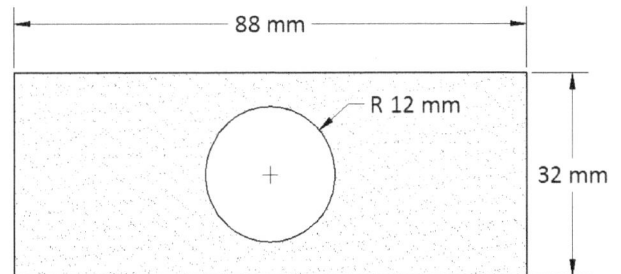

88 mm

R 12 mm

32 mm

25. The Manual of Steel Construction states the tolerance for weight variation is plus or minus 2.5%. A 28 foot length of structural tubing is designed to weigh 60.75 pounds per foot. Calculate its weight and the heaviest and lightest it can be inside of the 2.5% tolerance. Round your answer to the nearest pound.

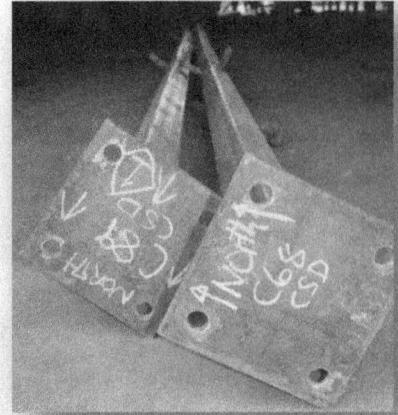

26. A stair stringer is to be cut with a ratio of rise to run of $7\frac{1}{8}"$ to $11\frac{1}{2}"$. If the total rise is $102\frac{3}{4}"$, calculate the total run rounded to the nearest 16th of an inch. Note: The picture is only to help understand the problem; the actual stringer will have more than nine steps.

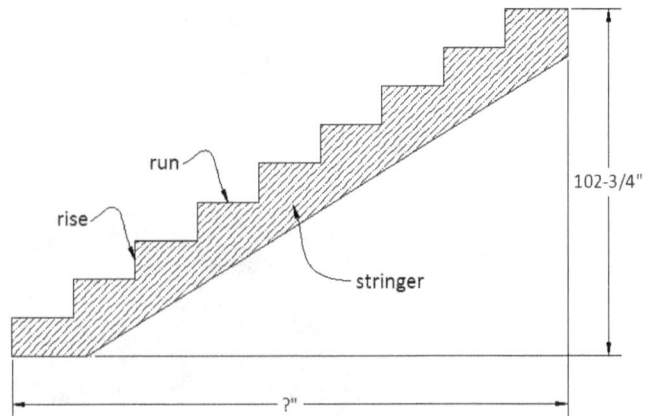

27. Resistors are common in electrical circuits and used to resist the flow of electricity. The colors identify the value of the resistor. Example:

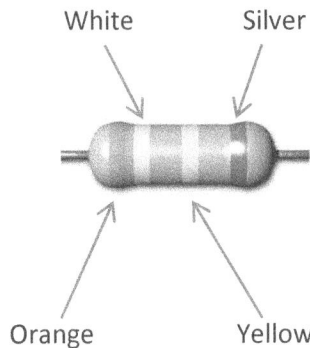

White   Silver

Orange   Yellow

Orange = 3, White = 9 and Yellow = 10^4.

Standard EIA Color Code Table 4 Band: ±2%, ±5%, and ±10%

Color	1st Band (1st figure)	2nd Band (2nd figure)	3rd Band (multiplier)	4th Band (tolerance)
Black	0	0	10^0	
Brown	1	1	10^1	
Red	2	2	10^2	±2%
Orange	3	3	10^3	
Yellow	4	4	10^4	
Green	5	5	10^5	
Blue	6	6	10^6	
Violet	7	7	10^7	
Gray	8	8	10^8	
White	9	9	10^9	
Gold			10^{-1}	±5%
Silver			10^{-2}	±10%

The design value of the resistor is 39 x 10^4 or 390000 Ω or 390 kΩ. The silver band represents a tolerance of ± 10%. 10% of 390 = 39, so the minimum value is 390 – 39 or 351 kΩ and the maximum value is 390 + 39 or 429 kΩ.

Find the design value, minimum value and maximum value for each of the resistors. Answer in kΩ.

a) 1st band = red, 2nd band = green, 3rd band = orange, 4th band = gold

b) 1st band = gray, 2nd band = blue, 3rd band = yellow, 4th band = red

c) 1st band = violet, 2nd band = white, 3rd band = red, 4th band = silver

28. Kirchhoff's Voltage Law (KVL) states that $V_t = V_1 + V_2 + V_3 + V_4 \ldots$ for a series circuit.
    a) Calculate the voltage at R_4 in the series circuit using KVL.
    b) Calculate the percentage of the total voltage at R_2.
    c) Calculate the percentage of the total voltage across R_1 and R_2 combined.

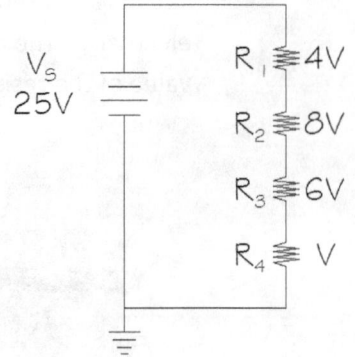

V_s
25V

$R_1 \lessgtr 4V$

$R_2 \lessgtr 8V$

$R_3 \lessgtr 6V$

$R_4 \lessgtr \;\;V$

29. Kirchhoff's Current Law (KCL) states that $I_t = I_1 + I_2 + I_3 + I_4 \ldots$ for a parallel circuit. Current is measured in amps but abbreviated with an I.

    a) Calculate the total current I_t in the parallel circuit using KCL.
    b) Calculate the percentage of the total current at R_3.
    c) Calculate the percentage of the total current across R_1 and R_2 combined.

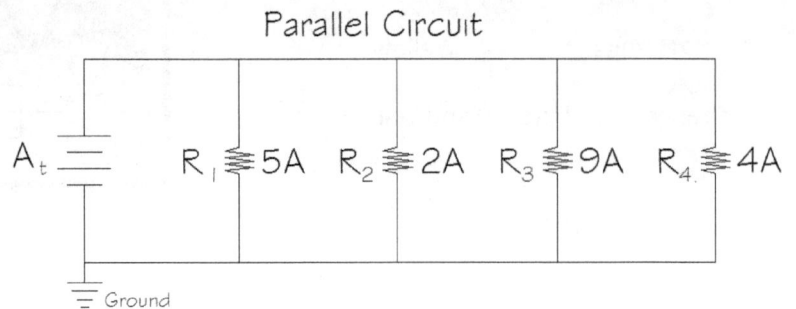

Parallel Circuit

A_t

$R_1 \lessgtr 5A \quad R_2 \lessgtr 2A \quad R_3 \lessgtr 9A \quad R_4 \lessgtr 4A$

Ground

30. Transformers are used in electronics to step up to a higher voltage or step down to a lower voltage. Step down transformers are often located at the top of telephone poles to reduce the overhead voltage to a level suitable for a home.

The Step Down Transformer diagram shows 1000 V on the primary side stepping down to 200 V on the secondary side. As the diagram illustrates, it is done with the number of turns or windings. The following ratio describes the relationship:

Step Down Transformer

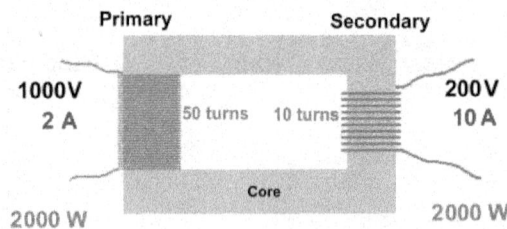

Primary   Secondary

1000V   200 V
2 A   50 turns 10 turns   10 A

Core

2000 W   2000 W

$$\frac{V_p}{V_s} = \frac{N_p}{N_s}$$

V_p = Voltage at the Primary side
V_s = Voltage at the Secondary side
N_p = Number of turns of wire at the Primary side
N_s = Number of turns of wire at the Secondary side

Calculate the missing values in the table. Round all values to the nearest whole number when necessary.

V_p	V_s	N_p	N_s
240 V	40 V	32	
320 V	24 V		14
65 V		25	4
	8 V	20	12

31. **Challenge Problem:** Taper is the difference between the diameters at each end of a part of a given length. A reamer is a tapered drill bit that can bore a hole of diameter C if inserted to depth D. Use a proportion to fill in the missing values in the chart accurate to 3 decimal places.

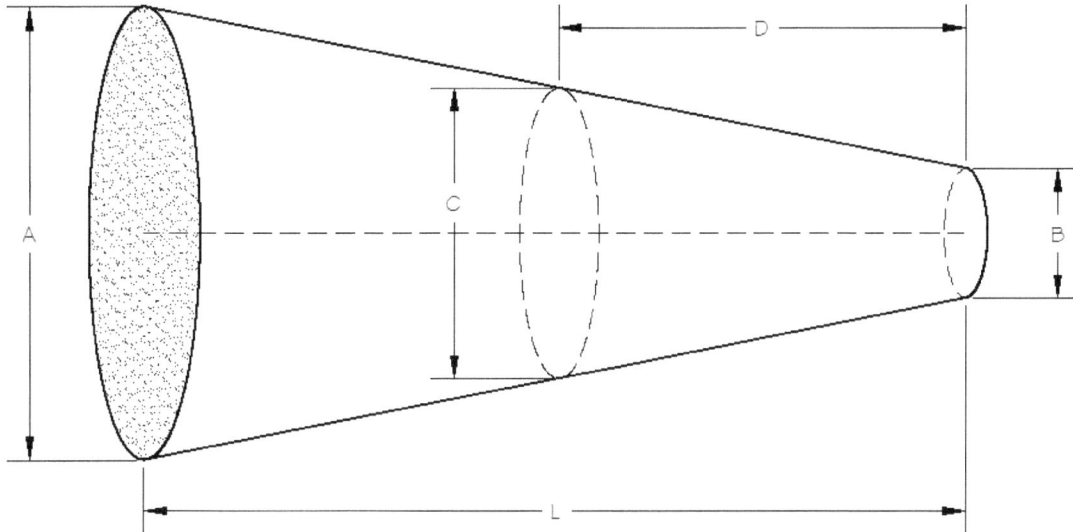

Length L	Diameter A	Diameter B	Depth D	Diameter C
4"	1.2"	.5"	1.4"	see example
5"	1.5"	.75"	2"	
84 mm	2.8 mm	1.2 mm		2 mm
9.5 cm	4.2 cm	3.2 cm		3.6 cm

Example:

$$\frac{Taper_1}{Length_1} = \frac{Taper_2}{Length_2}$$

$$\frac{1.2-.5}{4} = \frac{T_2}{1.4}$$

$$\frac{.7}{4} = \frac{T_2}{1.4}$$

1.4 x .7 = 4 x T_2

.98 = 4 x T_2

.245 = T_2

C - .5= .245 ... C = .745"

1.4: Dimensional Analysis

Do not get intimidated by the title of this section, the concept is simple. The units (dimensions) for a number will help, rather than add to, your mathematical difficulties.

Students often have trouble remembering if area, with dimensions measured in feet, is labeled ft, ft^2 or ft^3. Pay attention to the units of the numbers and there is nothing to remember. Finding the area of a rectangular concrete porch that is 8ft x 9ft is 72, since the area of a rectangle is L x W. Notice that if you pay attention to the units you are also multiplying ft x ft which equals ft^2. The answer is 72 ft^2.

A typical concrete problem involves calculating the amount to order for a sidewalk shaped like a rectangular box. Suppose it is 4 inches deep, 6 feet wide and 100 feet long. If you do not pay attention to the units you <u>cannot</u> order the correct amount. To make matters worse, concrete is ordered by the cubic yard (yd^3). Analyzing the dimensions or units of a number is indispensable in mathematics.

Let's put these two concepts together to solve the sidewalk problem:

Example 1.4.1: Calculating concrete volume

Find the volume of concrete in yd^3 needed for a sidewalk that is 4 in deep by 6 ft wide by 100 ft long.

Solution:

You may remember that the volume of a box is L x W x H (see the appendix for common formulas). If you multiply 4 in x 6 ft x 100 ft, you would get 2400, which is wrong, and the units would be ft^2-in which do not make sense.

A cubic inch (in^3) is not only a unit; it is literally a cube that is 1in x 1in x 1in. Start by converting the dimensions of the sidewalk to inches. Since there are 12 inches in 1 foot, let the units and some basic algebra skills do the work. $6ft \cdot \frac{12in}{1ft}$ would change the units of 6 ft without changing the actual length. You multiply fractions straight across so $\frac{6ft}{1} \cdot \frac{12in}{1ft} = \frac{6\cancel{ft}}{1} \cdot \frac{12in}{1\cancel{ft}} = 72in$.

The idea is this, if you need the units of feet to change to inches then you would have to multiply by the unit $\left(\frac{in}{ft}\right)$. Similarly, 100 ft = 1200 in, so we have 4 in x 72 in x 1200 in = 345,600 in^3.

This is the correct answer but needs to be in yd^3. We would have to multiply by the unit $\frac{yd^3}{in^3}$ to change to yd^3. The appendix at the end of this text lists all the conversions you will need for this section. Observe that 46,656 in^3 = 1 yd^3. Therefore the problem becomes:

$\frac{345,600in^3}{1} \cdot \frac{1yd^3}{46,656in^3}$ notice that you are actually multiplying by 1 since 46,656 in^3 = 1 yd^3

$\frac{345,600\cancel{in^3}}{1} \cdot \frac{1yd^3}{46,656\cancel{in^3}}$ cubic inches "cancel" out leaving cubic yards and we are led to divide

$\frac{345,600yd^3}{46,656} \approx 7.4 \; yd^3$ simplify

Final Answer: Volume ≈ 7.4 yd^3 of concrete.

This technique can also be used to change to different types of units like ft^3 to gallons, as the following example illustrates:

Example 1.4.2: Changing from cubic feet to gallons

Volumes of liquid are commonly measured in gallons. Find the number of gallons needed to fill a 4 ft x 9 ft x 14 ft box.

Solution:

The volume of the box would be 504 ft^3 using the formula from the previous example. Changing the units to gallons (gal) from ft^3 would require us to multiply by the unit $\left(\frac{\text{gal}}{\text{ft}^3}\right)$. The appendix gives the conversion: 1 ft^3 = 7.5 gal.

$$\frac{504\text{ft}^3}{1} \cdot \frac{7.5\text{ gal}}{1\text{ft}^3}$$

$$\frac{504\cancel{\text{ft}^3}}{1} \cdot \frac{7.5\text{ gal}}{1\cancel{\text{ft}^3}}$$ cubic feet cancel out leaving gallons and we are led to multiply

504×7.5 gal = 3780 gallons   simplify

Final Answer: A volume of 3780 gallons of water are needed to fill the box.

It may seem that multiplying 504 by 7.5 would change the answer, but remember that since 1 ft^3 = 7.5 gal that the fraction $\frac{7.5\text{ gal}}{1\text{ft}^3}$ actually equals 1. We are changing the units and the number without changing the actual amount of liquid.

This technique also helps convert back and forth from metric to standard units:

Example 1.4.3: Converting from metric to standard weight

A length of steel tubing is labeled to weigh 8700 grams, convert this weight to pounds to determine how much trouble it will be to lift.

Solution:

Changing the units to pounds (lbs) from grams (g) would require us to multiply by the unit $\left(\frac{\text{lbs}}{\text{g}}\right)$. The appendix gives the conversion: 1 lb = 453.6 g.

$$\frac{8700\text{g}}{1} \cdot \frac{1\text{ lb}}{453.6\text{ g}}$$

$$\frac{8700\cancel{\text{g}}}{1} \cdot \frac{1\text{ lb}}{453.6\cancel{\text{ g}}}$$ the grams cancel out leaving pounds and we are led to divide

$$\frac{8700\text{ lbs}}{453.6} \approx 19.2 \text{ pounds}$$ simplify

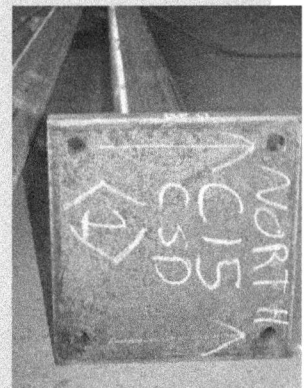

Final Answer: Weight ≈ 19.2 pounds, an easy size to work with, let's carry three at a time!

Consider an example where there are multiple units to change:

Example 1.4.4: Converting a metric speed to miles per hour (MPH)

Convert a speed of 24 meters/sec to mph ($\frac{miles}{hour}$).

Solution:

Start with 24 m/s and multiply by conversion factors that change seconds into hours and meters into miles:

$$\frac{24\ m}{1sec} \cdot \frac{60\ sec}{1\ min} \cdot \frac{60min}{1\ hr} \cdot \frac{3.281\ ft}{1\ m} \cdot \frac{1\ mile}{5280\ ft}$$ conversions taken from the appendix

$$\frac{24\ \cancel{m}}{1\ \cancel{sec}} \cdot \frac{60\ \cancel{sec}}{1\ \cancel{min}} \cdot \frac{60\ \cancel{min}}{1\ hr} \cdot \frac{3.281\ ft}{1\ \cancel{m}} \cdot \frac{1\ mile}{5280\ \cancel{ft}}$$ the remaining units are $\frac{miles}{hour}$

$$\frac{24*60*60*3.281\ mile}{5280\ hr} \approx 53.7\ mph$$ we know exactly what to multiply and divide

Final Answer: Speed ≈ 53.7 mph

Note: Here is another possibility: $\frac{24\ \cancel{m}}{1\cancel{sec}} \cdot \frac{3600\ \cancel{sec}}{1\ hr} \cdot \frac{100\ \cancel{cm}}{1\ \cancel{m}} \cdot \frac{1\ \cancel{in}}{2.54\ \cancel{cm}} \cdot \frac{1\ \cancel{ft}}{12\ \cancel{in}} \cdot \frac{1\ mile}{5280\ \cancel{ft}} \approx 53.7$ mph.

The key is that the units lead you to the math that is required. This problem would be daunting if left entirely to common sense, unless you are uncommonly sensible.

Section 1.4: Dimensional Analysis

Refer to the appendix for common conversions

Length Conversions

1. Convert 38 inches into a measurement in centimeters, rounded to one decimal place.

2. Convert 157 centimeters into a measurement in inches, rounded to the nearest 16th of an inch.

3. Convert 51 inches into a measurement in millimeters, rounded to one decimal place.

4. Convert 9 meters into a measurement in feet and inches, rounded to the nearest 16th of an inch.

5. Convert 9456 feet into a measurement in miles, rounded to two decimal places.

6. The measurements in the drawing are given in inches. Convert each dimension to centimeters, rounded to one decimal place.

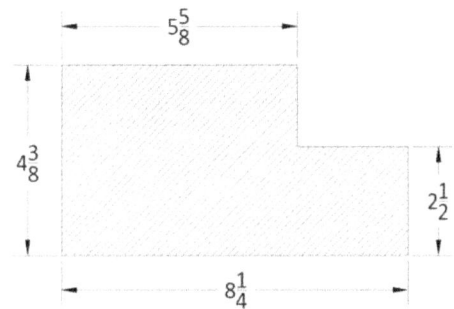

7. The measurements in the drawing are given in millimeters. Convert each dimension to inches rounded to the nearest 16th of an inch.

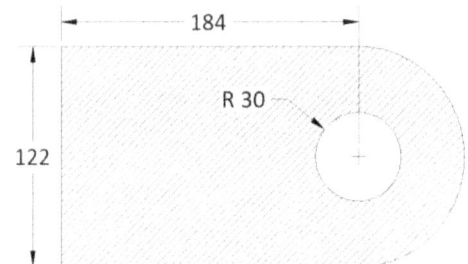

Area Conversions

8. Convert 17 square inches into a measurement in square centimeters, rounded to two decimal places.

9. Convert 236 square inches into a measurement in square feet, rounded to two decimal places.

10. A lot in a subdivision is listed as .32 acres, convert .32 acres to a measurement in square feet, rounded to the nearest square foot.

11. Convert 78 square feet into a measurement in square yards, rounded to two decimal places.

12. Convert 789 square inches into a measurement in square meters, rounded to two decimal places.

13. Convert five square meters into a measurement in square feet, rounded to two decimal places.

Volume Conversions

14. A 320 mL sample of blood is drawn. Convert it to ounces, rounded to one decimal place.

15. Convert five cubic feet into a measurement in gallons, rounded to two decimal places.

16. Convert 167 cubic feet into a measurement in cubic yards, rounded to two decimal places.

17. Convert a liquid measure of 4 cups to cm^3 (consider that there are 16 cups in a gallon or 8 ounces in a cup), rounded to the nearest cubic centimeter.

18. Convert four cubic feet into a measurement in cubic inches.

19. Convert 183,487 cubic inches into a measurement in cubic yards, rounded to one decimal place.

Rate Conversions

20. Convert 23 gallons per minute (GPM) into cubic feet per day, rounded to the nearest whole number.

21. A tree grows 3/8" per day. Convert this growth rate into feet per year, rounded to two decimal places.

22. A paint striper covers 100 meters in 28 seconds. Convert this speed to MPH, rounded to one decimal place.

23. An assembly line manufactures I-beams at a rate of 58 feet per second. Convert this speed to miles per hour (MPH), rounded to one decimal place.

24. A blood sample has a glucose level of $\frac{84mg}{dL}$. Convert this to $\frac{g}{L}$ (grams per liter).

25. Convert a rate of 18 feet per second into miles per hour (MPH), rounded to one decimal place.

26. Convert 27 miles per hour (MPH) into meters per second, rounded to one decimal place.

Weight Conversions

27. Use the chart of weights of common metals to make the conversions rounded to one decimal place:

    a. Convert the weight of Copper to ounces per cubic foot.

    b. Convert the weight of Tungsten to grams per cubic inch.

    c. Convert the weight of Magnesium to grams per cubic foot.

    d. Convert the weight of Stainless steel to pounds per cubic foot.

    e. Convert the weight of Titanium to grams per cubic centimeter.

Material	Weight (oz/in^3)
Magnesium	1.006
Aluminum	1.561
Titanium	2.641
Zinc	4.104
Stainless Steel	4.538
Copper	5.180
Tungsten	11.157
Gold	11.169

Weights of Common Metals

28. Find the weight of a five gallon bucket of concrete measured in pounds, rounded to one decimal place. Concrete weighs 3,915 pounds per cubic yard.

29. Find the weight of a four cubic foot wheel barrow filled with concrete measured in pounds, rounded to the nearest pound. Concrete weighs 3,915 pounds per cubic yard.

30. A grain is small unit of weight that pharmacists use (15 grains = 1 gram). Find the number of grains that should be measured into a 600 mg capsule.

31. Find the weight of a 16 cubic foot laminated veneer lumber (LVL) beam measured in pounds, rounded to the nearest pound. LVL beams weigh .34 ounces per cubic inch.

32. Find the weight of two 5 gallon buckets of water measured in pounds, rounded to one decimal place. Water weighs 62.4 pounds per cubic foot.

33. Find the weight of a 3,072 cubic inch Douglas fir header measured in pounds, rounded to one decimal place. Douglas fir weighs 38 pounds per cubic foot.

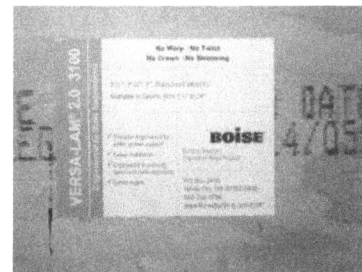

34. A nurse collected 920 mL of urine from a patient over a 10 hour period. The lab analysis showed it to contain a protein concentration of 40 $\frac{mg}{dL}$. Find the total amount of protein excreted by the patient. Find the rate of excretion in milligrams per day.

1.5: Order of Operations

We finish the chapter with the order of operations used without explanation in the introduction to this book. It is impossible to understand algebra in a meaningful way without a deep understanding of this foundational concept. The order is easily memorized with the acronym PEMDAS, which is often expanded to Please Excuse My Dear Aunt Sally, as a mnemonic device.

1. **P**arentheses
2. **E**xponents
3. **M**ultiply/**D**ivide
4. **A**dd/**S**ubtract

For me this is a bit uninspired for such an important concept so my personal favorite is: **P**eople **E**njoying **M**ath **D**isplay **A**wkward **S**ymptoms. It is acceptable to have a little fun while learning math, in fact, if you are not having fun you are doing something wrong. To deepen your grasp of this pivotal concept, let's take a minute and discuss the reason for this order.

I am embarrassed to admit that I graduated from college with a math degree without knowing what this reason is. Addition is the foundational math operation; it can be argued that addition is the only math operation. Consider the problem 3 x 4, you have memorized that it is 12 to save time. The answer is 12 because 3 x 4 means three-fours or 4 + 4 + 4. We invent multiplication to speed up the process when repeatedly adding the same number. For example, you buy 42 studs to build a wall costing $3.52 each. The total cost is found by adding $3.52 to itself 42 times. Multiplication thankfully speeds this process up, but the fact remains that multiplication is merely a useful technique for adding quickly.

If a formula requires you to find 3 x 6 + 5, understanding that it is shorthand for 6 + 6 + 6 + 5 which equals 23, implies that you must multiply first or you won't get it right. If you add before multiplying, cashier would charge you $33 instead of $23 for three $6 pipes and one $5 saw. The order does not result from my dear Aunt Sally. She became famous later merely for being a dear woman whose name started with an S. What does she have to say about exponents preceding multiplication?

Again the issue is resolved in the meaning of an exponent. 5^3 means 5 x 5 x 5. Exponents are shorthand for multiplication. Consider the expression $5 \times 4 + 3^2$.

5 x 4 + 3 x 3	using the definition of exponents
4 + 4 + 4 + 4 + 4 + 3 + 3 + 3	using the definition of multiplication
29	simplify (notice the complex expression is all addition when expanded to show its meaning)

Multiplication and exponents make it convenient for writing math expressions that would otherwise be very cumbersome, but the shorthand comes at the price of needing some skill to interpret it.

I have made no mention of subtraction and division. This is because they are simply invented as inverse operations. Division is just the inverse of multiplication and subtraction is the inverse of addition. It is important to understand that multiplication does not precede division; in fact, these operations must be performed from left to right as you would read a sentence. Similarly, addition and subtraction operations must also be performed working left to right.

Indeed then, in a sense, there is only one operation in math, addition. All other operations are either inverse operations or shortcuts. Although I would not recommend shouting this revelation the next time you are in a crowd for fear of being misunderstood, the fact should solidify your grasp on the order of operations. It should also drive home the fact that mathematics is not the memorization of useless and unrelated facts. It can be understood at a foundational level and applied to life in useful and time-saving ways.

Finally, parentheses are first in the order simply as a grouping symbol with the express meaning "do this first". It must be noted that large division bars in an expression like $\frac{3+4\times6}{5^2-7}$ imply parentheses, so that you must simplify the top (numerator) and the bottom (denominator) before performing the division. Most calculators allow for entering an entire expression like this, but do not have a large division symbol and so must be entered as $(3 + 4 * 6)/(5\char`\^2 - 7)$. It must also be noted that large square root symbols carry the same implication. An expression like $\sqrt{1 + 4^2}$ requires that the inside of the square root be determined as 17 before taking the square root. Again you must enter $\sqrt{(1 + 4\char`\^2}$) into a calculator to get the correct answer since there is no large square root symbol.

Consider a simple example involving money growing at an interest rate:

Example 1.5.1: Calculating simple interest

Solve for the final amount of money (A), if the principal (P) = $460, the interest rate (r) = 8.3%, and the time (t) = 10 years. Use the formula: $A = P + Prt$.

Solution:

$A = 460 + 460 * .083 * 10$	substitute the numbers into the formula
$A = 460 + 381.8$	multiplication is before addition and is done from left to right
$A = 841.8$	simplify

Final Answer: The amount is $841.80.

Consider an example involving the cornering force that a vehicle exerts on its passengers as a function of the radius of the turn and the speed of the vehicle:

Example 1.5.2: Calculating the force experienced when cornering in a vehicle

Solve for the force (F) measured in g's, if you know the velocity (v) = 48 mph and the radius (r) = 24 feet. Use the formula: $F = \frac{v^2}{14.957r}$. Round to the nearest tenth.

Solution:

$F = \frac{48^2}{14.957*24}$ substitute the numbers into the formula

$F = \frac{2304}{14.957*24}$ exponents are first, (take time here to learn your calculator, enter: 48 ^ 2 or 48 x^y 2, depending on the type of calculator that you have, this will save time when the exponent is larger than 2)

$F = \frac{2304}{358.968}$ remember the large division implies a parentheses so the top and bottom must be simplified first (if you enter 2304 / 14.957 x 24 in your calculator it will be wrong)

$F \approx 6.4$ simplify

Final Answer: The cornering force is F ≈ 6.4 g's, meaning about six and half times the force of gravity.

Consider a more realistic and complex example of money growing at a monthly interest rate:

Example 1.5.3: Compound interest formula

Solve for the amount (A) if the principal (P) = $580, the interest rate (r) = 7.2%, and the time (M) = 42 months. Use the formula: $A = P(1 + \frac{r}{12})^M$.

Solution:

$A = 580\left(1 + \frac{.072}{12}\right)^{42}$ substitute the numbers into the formula

$A = 580 * 1.006^{42}$ parentheses are first and $1 + \frac{.072}{12}$ simplifies to 1.006 when dividing before adding

$A \approx 580 * 1.286$ exponents are next, 1.006^{42} simplifies to 1.286 (rounded to the thousandth place and I hope you figured out how to enter this in your calculator)

$A \approx 745.88$ simplify

Final Answer: A deposit of $580 at 7.2% APR for 42 months would grow to $745.88.

Note: A bank would report the amount as $745.66 since they would not round until the end.

Consider a formula used to calculate roofing for a house, involving a large square root sign:

Example 1.5.4: Calculating square feet of roofing

Solve for the number of square feet of roofing (R), if the slope of the roof (s) is $\frac{5}{12}$ and the floor area (A) = 1860 ft^2. Use the formula: $R = A\sqrt{1 + s^2}$. Round to the nearest square foot.

Solution:

$R = 1860\sqrt{1 + \left(\frac{5}{12}\right)^2}$ substitute the numbers into the formula

$R \approx 1860\sqrt{1 + (.417)^2}$ parentheses first and rounding to the thousandth place

$R \approx 1860\sqrt{1.174}$ The large square root implies a parenthesis and so $1 + .417^2$ must be simplified. Squaring .417 gives .174. Adding 1 results in 1.174.

$R \approx 1860 * 1.08$ a square root is actually an exponent of .5, so the square root is before multiplication in the order of operations

$R \approx 2009$ simplify

Final Answer: R ≈ 2009 square feet of roofing. Note: You will get R = 2015 if you do not round until the end. Rounding multiple times in a calculation can result in an answer that is remarkably inaccurate. This formula is a remarkable simplification of the difficult process it would be to calculate all the areas of each rectangle, triangle, parallelogram, and trapezoid that often compose a roof. A person who knows algebra can get more accurate material calculations and save a lot of time in the process.

For the last example we consider a remarkably complicated engineering formula for the deflection (bending) of a cantilevered beam. I will spare you the meaning of each letter and the complexity involved with more realistic numbers.

Example 1.5.5: Calculating beam deflection

Solve for D if w = 4, n = 2, L = 3, E = 5, and I = 6. Use the formula:
$D = \frac{w(n^4 - 4nL^3 + 3L^4)}{24EI}$.

Solution:

$D = \frac{4(2^4 - 4*2*3^3 + 3*3^4)}{24*5*6}$ substitute the numbers into the formula

$D = \frac{4(16 - 4*2*27 + 3*81)}{24*5*6}$ exponents in the parentheses first

$D = \frac{4(16 - 216 + 243)}{24*5*6}$ multiplications inside the parenthesis next

$D = \frac{4*43}{24*5*6}$ finish the parenthesis working left to right

$D = \frac{172}{720}$ since the large division sign implies a separate parentheses top and bottom, simplify the top and bottom separately

$D \approx .24$ simplify

Final Answer: D ≈ .24; the beam will bend or deflect approximately $\frac{1}{4}$ inch.

Section 1.5: Order of Operations

1. The volume (V) of concrete for a driveway or sidewalk is often estimated by $V = \frac{LW}{80}$.

    Where V = volume measured in cubic yards, L = length measured in feet, and W = width measured in feet.

    Find the volume of concrete needed for a driveway that is 24' wide and 12' long.

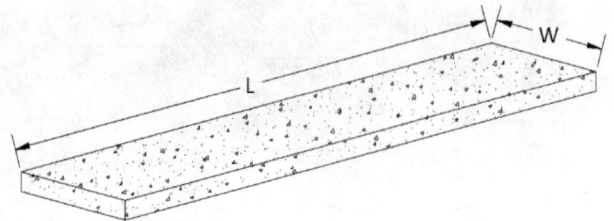

2. The volume (V) of concrete for a driveway or sidewalk is often estimated by $V = \frac{LW}{80}$.

    Where V = volume measured in cubic yards, L = length measured in feet, and W = width measured in feet.

    Find the volume of concrete needed for a sidewalk that is 5 feet wide and 124 feet long.

3. An approximation of the belt length (L) in a motor is $L = \pi(R + r) + 2d$.

    R = radius of the larger pulley, r = radius of the smaller pulley, and d = distance between the pulleys. Pi (π) ≈ 3.14 but most calculators have a pi button that should be used instead since it is quicker and more accurate.

    Find the length of the belt rounded to one decimal place, if R = 16 cm, r = 7 cm, d = 28 cm.

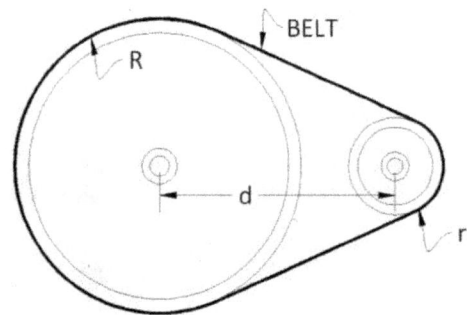

4. The carburetor size formula is $C = \frac{dRV}{3456}$.

    C = cubic flow modification (CFM), d = piston displacement, R = engine revolutions per minute (RPM), and V = volumetric efficiency.

    Find the CFM for a 3300 RPM engine with a 5.2-inch piston displacement and a volumetric efficiency of 124%, rounded to one decimal place.

5. The torque formula is $T = \dfrac{5252H}{R}$.

    T = torque, H = horsepower, and R = engine revolutions per minute (RPM).

    Find the torque for a 2300 RPM engine with 840 horsepower, rounded to one decimal place.

6. In electronics, Power (P) is $P = RI^2$.

    P = power measured in watts, R = resistance measured in ohms, and I = current measured in amps.

    Find the power consumed by an 8-ohm resistor with a 9-amp current passing through it.

7. The volume of cylindrical footing (V) is $V = \pi r^2 h$.

    H = height and r = radius.

    Find the volume rounded to one decimal place, if h = 14" and r = 8".

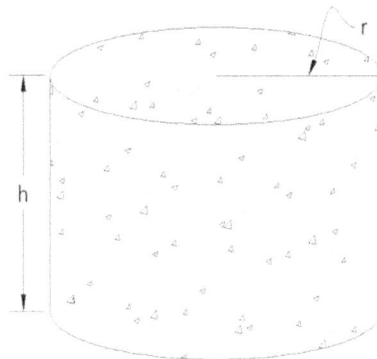

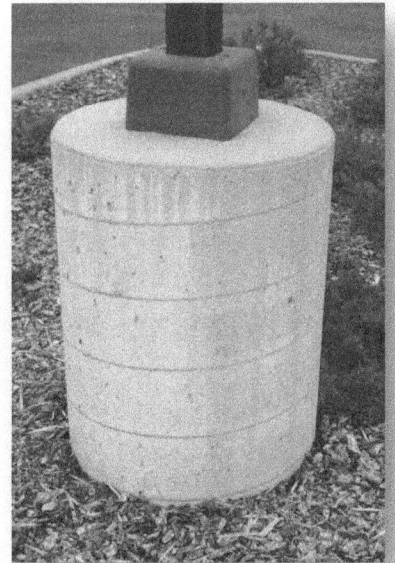

8. The area of a regular octagonal window (A) is $A = 4.828n^2$.

    Find the area of an octagonal window with a side of 12" rounded to one decimal place.

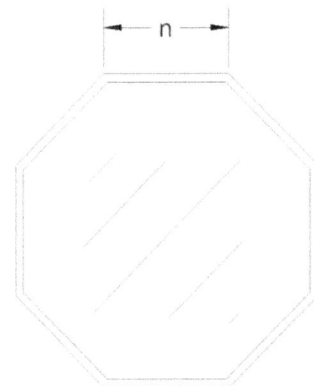

9. The moment of inertia (I) of a beam is $I = \dfrac{bd^3}{12}$.

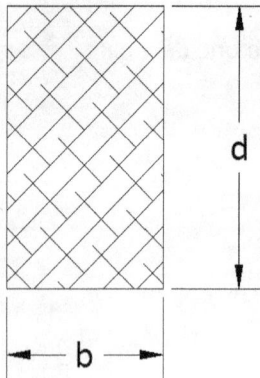

Note: Moment of inertia is a measure of a beam's effectiveness at resisting bending based on its cross-sectional shape.

I = moment of inertia of the beam measured in inches^4, b = width of the beam measured in inches and d = height of the beam measured in inches.

Find the moment of inertia of a beam rounded to one decimal place, if b = 3.5" and d = 14".

10. The speed of a car is $S = \dfrac{DR}{336G}$.

S = speed in miles per hour (MPH), D = tire diameter in inches, R = engine revolutions per minute (RPM), and G = gear ratio.

Find the speed of a car with 26 inch diameter tires, a 3400 RPM engine, and a gear ratio of 3.5, rounded to the nearest MPH.

11. The reactance offered by a capacitor in electronics is $X = \dfrac{1}{2\pi fC}$.

X = reactance measured in ohms, f = frequency measured in cycles per second (hertz), C = capacitor size measured in farads.

Find the reactance for a capacitor in a circuit with a frequency of 60 hertz and a capacitor size of .00012 farads, rounded to three decimal places.

12. The voltage drop in an electrical wire is $V = \dfrac{2LIR}{1000}$.

AWG	R
16	4.884
14	3.072
12	1.932
10	1.215
8	.764
6	.481
4	.302

V = voltage drop measured in volts, L = length of the wire measured in feet, I = current measured in amps and R = resistance in the wire measured in ohms.

Use the table at the right to determine the voltage drop in a 350 foot #12 AWG electrical cord attached to a saw drawing 13 amps of current. Round to one decimal place.

Recalculate if the cord size is increased to #10 AWG.

13. The point load deflection (D) of a beam is $D = \dfrac{PL^3}{48EI}$.

Note: Deflection is simply a measurement of the amount of bend in a beam.

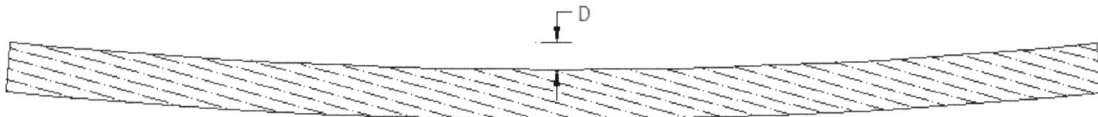

D = deflection measured in inches, P = weight on the beam measured in pounds, L = length of the beam measured in inches, E = elasticity of the beam measured in pounds per square inch (PSI), and I = moment of inertia of the beam measured in inches^4.

Find the deflection of a beam rounded to one decimal place if L = 216, P = 4500, E = 1,800,000, and I = 432.

14. The uniform load deflection (D) of a beam is $D = \dfrac{5PL^4}{384EI}$.

Note: Deflection is simply a measurement of the amount of bend in a beam.

D = deflection measured in inches, P = weight on the beam measured in pounds, L = length of the beam measured in inches, E = elasticity of the beam measured in pounds per square inch (PSI), and I = moment of inertia of the beam measured in inches^4.

Find the deflection of a beam rounded to one decimal place if L = 168 inches, P = 358 pounds, E = 2,000,000 psi, and I = 968 inches^4.

15. The formula to calculate the radius (R) of an arch window is $R = \dfrac{W^2 + 4H^2}{8H}$.

W = width of the window & H = height of the window.

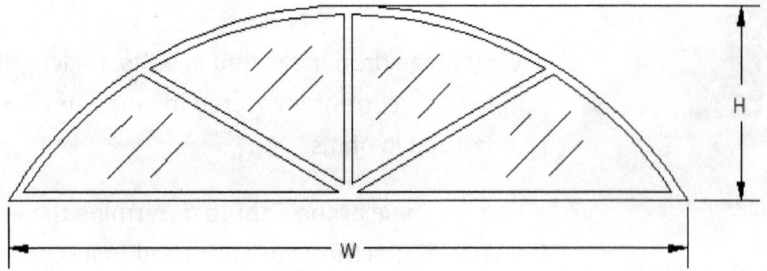

Find the radius of an arch window accurate to the 16th of an inch, which has a width of 42 inches and a height of 12 inches.

16. The formula to calculate size (S in inches) of a square footing is $S = 12\sqrt{\dfrac{W}{B}}$.

W = weight on the footing (in pounds) & B = soil bearing capacity in pounds per square foot (PSF).

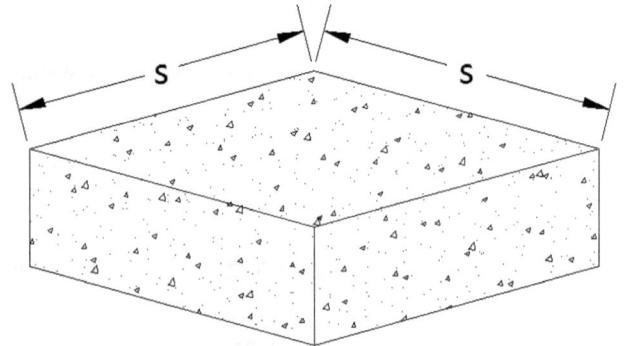

Find the size (S) of footing necessary to hold a weight of 7400 pounds that sits on soil able to bear 1500 psf. Round your answer to one decimal place.

17. The length of a rafter (R) can be calculated using the formula: $R = \frac{W}{2}\sqrt{S^2 + 1}$.

R = length of the rafter measured in inches, W = width of the building measured in inches, and S = slope of the roof.

Find the length of a rafter for a building that is 312" wide and has a slope of 7/12, rounded to the nearest 16th of an inch.

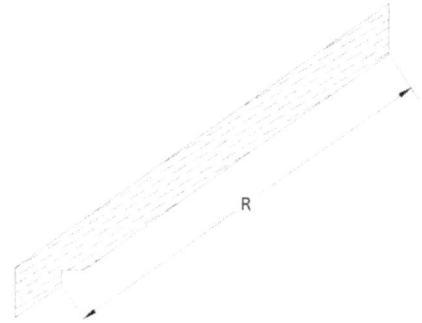

18. The allowable stress (S) on a post is $S = \frac{3ED^2}{10L^2}$.

S = allowable stress measured in pounds per square inch (PSI), D = dimension of the post measured in inches, L = length of the post measured in inches, and E = elasticity of the beam measured in pounds per square inch (PSI).

Find the allowable stress on a post rounded to one decimal place, if L = 120 inches, E = 1,100,000 psi, D = 3.5 inches.

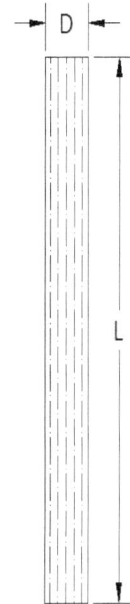

19. The Cornering Force (F) that a vehicle exerts on its passengers is $F = \frac{v^2}{32r}$.

F is measured in g's (one g is the force of gravity of the earth), v = velocity measured in feet per second and r = radius measured in feet.

Find the cornering force of a car traveling 82 feet per second around a corner of radius 48 feet, rounded to one decimal place.

20. The Engine Displacement (D) is $D = \frac{\pi b^2 sc}{4}$.

D = engine displacement measured in cubic centimeters, b = bore (diameter of the cylinder) measured in centimeters, s = stroke (distance that the piston travels) measured in centimeters, and c = number of cylinders.

Find the displacement of an 8 cylinder engine with a 2.8-cm bore and a 4.4-cm stroke, rounded to one decimal place.

21. The Exhaust Header Tubing Length (L) is $L = \frac{1900D}{d^2 R}$.

L = length measured in inches, D = displacement measured in cubic inches, d = exhaust head diameter measured in inches, and R = revolutions per minute (RPM).

Find L, if D = 350 in^3, d = 3 in, and R = 2200 RPM's, rounded to the nearest inch.

22. In electronics, Power (P) is $P = \frac{E^2}{R}$.

P = power measured in watts, R = resistance measured in ohms, and E = voltage measured in volts.

Find the power consumed by 120 volt electricity passing through a circuit with .6 ohms of resistance.

23. Roofing for a house can be ordered by the square (1 square = 100ft^2). The formula for calculating the number of squares of roofing for a house is $R = \frac{A\sqrt{1+S^2}}{100}$.

R = number of squares, A = area or square footage of the floor of the house, and S = slope of the roof. Find the number of squares of roofing for a 1950ft^2 house with a roof slope of 9/12, rounded up to the nearest whole number.

24. The formula for horsepower is $H = W\left(\frac{S}{234}\right)^3$.

H = horsepower, W = weight in pounds, and S = speed in MPH.

Calculate the horsepower for a car that weighs 2740 pounds and is capable of 106 miles per hour (MPH), rounded to the nearest horse. Horses, as you know, are weak with decimals.

25. The formula for resistance in a parallel electrical circuit is $R_t = \dfrac{R_1 R_2}{R_1 + R_2}$.

    R_t = total resistance, R_1= resistance one, and R_2= resistance two.

    Find the total resistance if R_1 = 460 Ω and R_2 = 720 Ω, rounded to one decimal place.
    Note: Ω is an electrical symbol for ohms.

26. The formula for speed is $S = 234 \left(\dfrac{H}{W}\right)^{.333}$.

    H = horsepower, W = weight in pounds, and S = speed in MPH.

    Calculate the speed for a car that weighs 2160 pounds with 712 horse power, rounded to the nearest MPH.

27. The formula to calculate impedance in an RL circuit is: $Z = \sqrt{R^2 + X^2}$.
    Z = impedance measured in ohms
    R = resistance measured in ohms
    X = reactance measured in ohms

    Calculate the impedance in a circuit with 2.3 kΩ of resistance and a reactance of 5.4 kΩ, rounded to the nearest tenth of a kΩ.

28. The formula to calculate inductive reactance is: $X = 2\pi f L$
    X = inductive reactance measured in ohms
    f = frequency measured in hertz (Hz)
    L = inductance measured in henrys (H)

    Calculate the inductive reactance for an AC circuit with a frequency of 3 kHz and an inductance of 12.3 mH. Note: kHz and mH have metric prefixes that must be considered. Round to the nearest whole number.

29. Use the formula to convert a patient temperature of 36° Celsius to Fahrenheit.
    F = 1.8C + 32

30. Use the Voltage Divider Formulas to calculate voltages V_1 and V_2 at each resistor R_1 and R_2.

V_1 = voltage at R_1

V_2 = voltage at R_2

R_1 = resistance at R_1 measured in ohms

R_2 = resistance at R_2 measured in ohms

R_t = total resistance measured in ohms

$$V_1 = V_t\left(\frac{R_1}{R_t}\right) \quad \& \quad V_2 = V_t\left(\frac{R_2}{R_t}\right) \quad \text{also note that}$$

$R_t = R_1 + R_2$

Series Circuit

V_t
20V

$R_1 \lessgtr 120\,\Omega$

$R_2 \lessgtr 680\,\Omega$

Ground

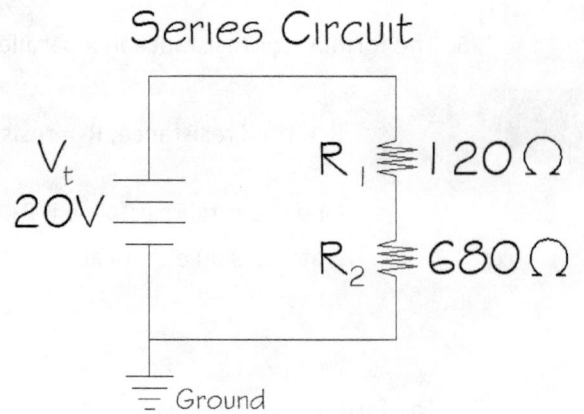

31. Use the Current Divider Formulas to calculate currents I_1 and I_2 at each resistor R_1 and R_2.

Answer in amps rounded to 2 decimal places.

I_1 = amperage at R_1

I_2 = amperage at R_2

R_1 = resistance at R_1 measured in ohms

R_2 = resistance at R_2 measured in ohms

R_t = total resistance measured in ohms

Parallel Circuit

V_t
10V

$R_1 \lessgtr 150\,\Omega$ $R_2 \lessgtr 270\,\Omega$

Ground

$$I_1 = I_t\left(\frac{R_t}{R_1}\right) \quad \& \quad I_2 = I_t\left(\frac{R_t}{R_2}\right) \quad \text{also note that } R_t = \frac{R_1 R_2}{R_1 + R_2} \text{ and } I_t = \frac{V_t}{R_t}$$

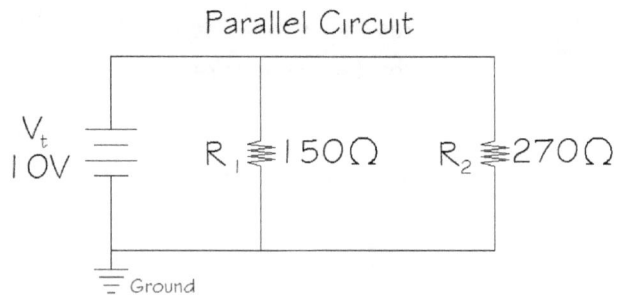

32. **Challenge Problem:** the moment of inertia (I) of an I-joist is $I = \frac{bd^3 - (b-c)(d-2a)^3}{12}$.

Note: Moment of inertia is a measure of a beam's effectiveness at resisting bending based on its cross-sectional shape.

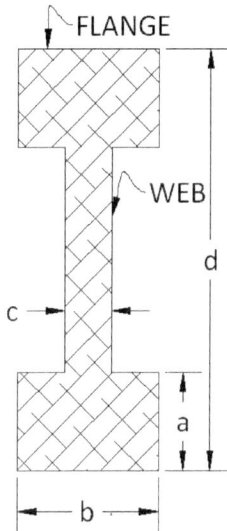

I = moment of inertia of the beam measured in inches^4, a = flange thickness measured in inches, b = width of the flange measured in inches, c = web thickness measured in inches, and d = height of the joist measured in inches.

Find the moment of inertia of Boise Cascade's 5000 series BCI where a = 2-1/2",

b = 1-1/8", c = 3/8" and d = 9-1/2". Round your answer to one decimal place.

33. **Challenge Problem:** A keyway is often cut in a cylindrical shaft in machining to lock two parts together. Achieving the desired depth (k) can be accomplished by calculating the depth of cut (d). The formula for calculating the depth (d) of the cut is $d = \frac{2k + D - \sqrt{D^2 - w^2}}{2}$.

Find d if k = .250, D = 2.875, and w = .625 (all measures are in decimal inches). Round your answer to the nearest thousandth of an inch.

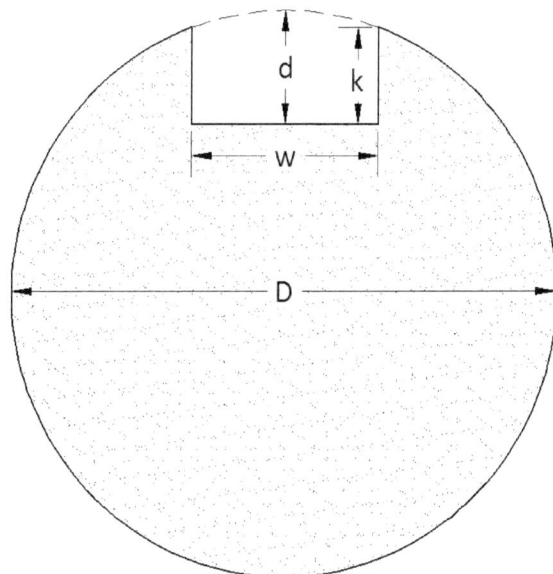

Chapter 2:
Formulas/Equations

In chapter 1 we limited our use of formulas to solving for the letter that is alone, typically on the left side of an equation.

As the following example will illustrate, it is often useful to solve for another letter in a formula.

Recall the formula used to calculate the point load deflection of a beam from section 1.5:

$$D = \frac{PL^3}{48EI}.$$

Where P = weight on the beam (in pounds), L = length of the beam (in inches), E = elasticity of the beam (in psi) & I = moment of inertia of the beam (in inches^4).

The most practical use of this formula actually involves solving for I not D. A carpenter could use the formula to calculate D, but if he discovered that the deflection of the beam was too great, he would have to change one of the other values to compensate. The deflection of a beam cannot exceed 1" in a residential application, and the most practical value to change is the size of the beam which involves changing I.

Example 2.1: Beam Deflection

Solve for I, if D = 1", P = 900 lbs., L = 132", and E = 1,200,000 psi (pounds per square inch). Use the formula: $D = \frac{PL^3}{48EI}$.

Solution:

$1 = \dfrac{900*132^3}{48*1,200,000*I}$ enter the numbers into the formula

$1.8 \approx \dfrac{900*132^3}{48*1,200,000*20}$ guessing a value of 20 for I

$1.2 \approx \dfrac{900*132^3}{48*1,200,000*30}$ guessing a value of 30 for I

$.9 \approx \dfrac{900*132^3}{48*1,200,000*40}$ guessing a value of 40 for I

Final Answer: Guessing and checking: I ≈ 36 in⁴. Choosing a beam with an I-value of 36 or higher would keep the deflection under 1".

In life, you often know the result that you are after, the problem is figuring out what will lead to this result. I don't want a formula that calculates my weight: Weight = body type + calories – exercise. I want to know what to eat and how much to exercise will result in my desired weight. The purpose of chapter 2 is to learn how to solve for letters buried in formulas without having to guess and check. It is a waste of time and you usually settle for an approximate answer.

2.1: Solving simple equations

In this section we will learn how to solve for I in the beam deflection formula without guessing at all. A simple illustration for solving equations is found in wrapping a present. The order of operations for wrapping a present is:

1. Put it in a box
2. Wrap it with paper
3. Tie a ribbon around it
4. Put a bow on it

When you unwrap a present it is done in the reverse order. You have to remove the bow, then the ribbon, next the paper and finally the box to get at the present. Not only is the order of removal reversed but even the process at each stage is reversed (i.e. tie and untie, wrap and unwrap). In the beam deflection formula the letter I represents the present hidden inside its wrapping (admittedly, the illustration is a bit strained since the letter I is a lousy present). Recall the four step order of operations that we studied in section 1.5:

1. **P**arentheses
2. **E**xponents
3. **M**ultiply/**D**ivide
4. **A**dd/**S**ubtract

In this section we are going to "unwrap" the letter we are solving for by removing the numbers around it in the reverse order. When solving for a letter that is surrounded by numbers and operations, "unwrap" it by first removing any numbers that are added or subtracted, then remove numbers that are multiplied or divided, next remove exponents and finally any numbers inside parentheses. As with a present, if any of these four operations are missing then skip it and move on. My wife notes that my presents only have paper and a bow. "That is so they are easier for you to unwrap", I lovingly respond. Most equations will be missing at least one of the four operations, just as most presents will be missing at least one wrapping (usually the ribbon since the sliding scissor trick rarely results in an attractive spiral). In this section we will limit our problems to those involving only add, subtract, multiply and divide, saving parentheses and exponents for section 2.2.

Consider an example involving money growing at an interest rate:

Example 2.1.1: Calculating a simple interest rate

Solve for the interest rate (r) if the amount (A) = $222, time (t) = 6 years, and principal (P) = $150. Use the formula: $A = P + Prt$.

Solution:

$222 = 150 + 150 * r * 6$	substitute the numbers into the formula
$72 = 150 * r * 6$	150 is added to $150 * r * 6$, and the inverse operation for add is subtract, so subtract 150 from both sides
$12 = 150 * r$	150 & 6 are both multiplied by r so each can be removed by dividing, divide both sides by 6
r = .08	divide both sides by 150

Note: $150 * r * 6$ could be simplified to $900 * r$, before beginning to unwrap the r, since multiplication can be done in any order. Then you would only have to divide both sides by 900, $(72 \div 900 = .08)$.

Final Answer: You would need to get an 8% interest rate for $150 to grow to $222 in 6 years.

Consider a fencing example:

Example 2.1.2: Calculating fencing

The perimeter of a rectangular fence is given by the formula: $P = 2L + 2W$. Calculate the length (L) of a fence that could be built with 248 feet of fencing that is to have a width (W) of 38 feet.

Solution:

248 = 2L + 2•38	substitute the numbers into the formula
248 = 2L + 76	simplify
172 = 2L	subtract 76 from both sides (undo additions or subtractions first)
86 = L	divide both sides by 2 (undo multiplications or divisions next)

Final Answer: If the rectangular fence is built with a length of 86 feet and a width of 38 feet it will have a perimeter of 248 feet and use all of the fencing.

Let's try an example that requires solving for a letter in the denominator:

Example 2.1.3: Engine capacity

The engine torque formula is $T = \frac{5252H}{R}$. Find the RPM's (R) for an engine with 540 foot pounds of torque (T) and 352 horsepower (H), rounded to the nearest whole number.

Solution:

$540 = \frac{5252*352}{R}$	substitute the numbers into the formula
$540 = \frac{1,848,704}{R}$	simplify
$540R = 1,848,704$	multiply both sides by R
$R \approx 3424$	divide both sides by 540

Final Answer: The engine will be capable of 3424 RPM's.

Section 2.1: Solving Simple Equations

1. Headers for a door are always five inches larger than the door $H = D + 5$. Find the door size for a 41 inch header.

2. The weight (W) of a plastic water tank is modeled by $W = 8.345G + 67$.

   W = total weight of the water tank measured in pounds, G is the size of the water in the tank measured in gallons, and 67 pounds is the weight of the empty tank.

   Find the number of gallons that can be hauled by a truck capable of holding one ton (2000 pounds), rounded to the nearest gallon.

   Water & Liquid Storage Tanks => PLASTIC STORAGE TANKS

   Plastic Storage Tank

Part Number:	NW250VERT
Capacity:	250 Gallon Vertical Poly Storage Tank
Size:	30" dia. x 89"H
USD Price:	249.99
USD Shipping:	CALL FOR PRICING

   Contact Us .

   Add To Cart

   Plastic Storage Tanks

   8" Manway & 2" Drain Fitting

   Tanks are translucent white with gallon markers

   wt. 67 lbs

3. Beer's Law concerns radiation absorbance: $A = EcL$

   A = absorbance

   E = constant related to the material

   c = concentration

   L = path length

   Calculate the constant for an absorbance of 12.5, a concentration of 3.2, and a path length of 7.8. Round to 1 decimal place.

4. In design, the degrees of spacing (D) is modeled by $D = \dfrac{360}{n}$.

   D = degrees of spacing between each hole, 360 = degrees in a circle, and n = number of holes.

   Find the number of holes that can be drilled if they are 15° apart.

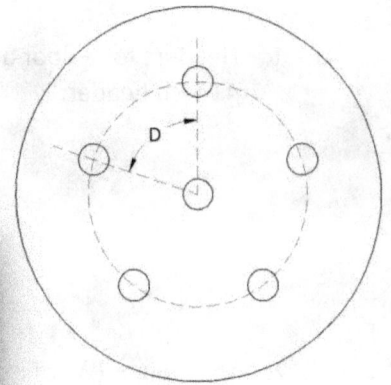

5. The weight (W) of a metal water tank is modeled by W = 8.345G + 3250.

   W = total weight of the water tank measured in pounds, G is the amount of water in the tank measured in gallons, and 3250 pounds is the weight of the empty tank.

   Find the number of gallons that can be hoisted by a crane capable of lifting 30 tons (60,000 pounds), rounded to the nearest whole number.

6. The Cost of a rental van (C) is modeled by: C = .32M + 37

   C = total cost, M is the number of miles driven at 32 cents per mile, and $37 = price of the rental.

   Find the number of miles that can be driven on a budget of $350, rounded to the nearest mile.

7. Use the formula to convert a patient temperature of 96° Fahrenheit to Celsius. Round to 1 decimal place.

   F = 1.8C + 32

8. A contractor bids a job at $680 for materials plus $42 per hour for labor. The total cost for the job can be modeled by: C = 42H + 680.

    Find the number of hours that he has for the job if the owner would like to total cost to be under $2000, rounded to the nearest hour.

9. The thickness (T) of the wall of a pipe can be found using the formula: $T = \dfrac{O.D. - I.D.}{2}$.

    Find the outside diameter if T = 1.32 mm and I.D. = 14 mm.

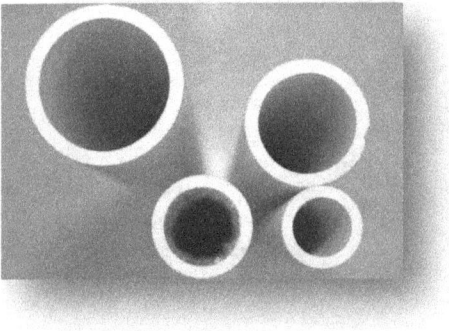

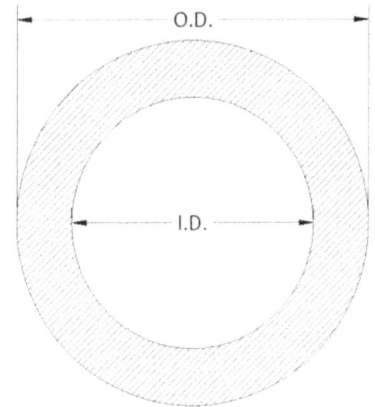

10. The thickness (T) of the wall of a pipe can be found using the formula: $T = \dfrac{O.D. - I.D.}{2}$.

    Find the inside diameter if T = 1/8"   O.D. = 2-3/8"

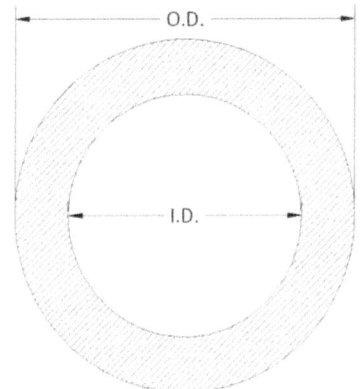

11. The torque formula is $T = \dfrac{5252H}{R}$.

    T = torque, H = horsepower, and R = engine revolutions per minute (RPM).

    Find the horsepower for a 2560 RPM engine with 680 foot pounds of torque, rounded to the nearest whole number.

12. The formula for the speed of a car is $S = \dfrac{DR}{336G}$.

    S = speed of the car in miles per hour (MPH), D = tire diameter, R = engine speed in revolutions per minute (RPM), and G = gear ratio.

    Calculate the gear ratio necessary to achieve a speed of 142 MPH in a car with 30 inch diameter tires and an engine speed of 4000 RPM's, rounded to the one decimal place.

13. The shear for a simple beam, uniformly loaded, can be found using the formula $V = w\left(\dfrac{L}{2} - x\right)$.

    Find length (L) measured in feet, if shear (V) = 438 ft-lbs, weight (w) = 47 lbs, and location (x) = 8 feet, rounded to one decimal place.

14. The formula for engine displacement is $D = \dfrac{\pi b^2 s c}{4}$.

    D = engine displacement, b = bore (diameter of the cylinder), s = stroke (distance that the piston travels), and c = number of cylinders.

    Calculate the stroke for a 6-cylinder engine with 350 cubic inches of displacement and a 4-inch bore, rounded to one decimal place.

15. The formula for horsepower is $H = W\left(\dfrac{S}{234}\right)^3$.

    H = horsepower, W = weight in pounds, and S = speed in MPH.

    Calculate the weight of a car with 480 horsepower and a speed of 132 miles per hour, rounded to the nearest pound.

16. The formula for the speed of a car is $S = \dfrac{DR}{336G}$.

    S = speed of the car in miles per hour (MPH), D = tire diameter, R = engine speed in revolutions per minute (RPM), and G = gear ratio.

    Calculate the engine RPM's necessary to achieve a speed of 135 MPH in a car with 28 inch diameter tires and a gear ratio of 3.8, rounded to the nearest RPM.

17. The piston speed formula is $P = \dfrac{sR}{6}$.

> P = piston speed measured in feet per minute (FPM), s = stroke length in inches, and R = engine speed in revolutions per minute (RPM).
>
> Calculate the RPM's for an engine with a 4.1-inch stroke and a piston speed of 1800 FPM, rounded to the nearest RPM.

18. The carburetor size formula is $C = \dfrac{dRV}{3456}$.

> C = cubic flow modification (CFM), d = piston displacement, R = engine revolutions per minute (RPM), and V = volumetric efficiency (%).
>
> Find the volumetric efficiency for a 4200 RPM engine with a 4.8-inch piston displacement and a CFM of 7.23. Answer as a percent rounded to one decimal place.

19. The torque formula is $T = \dfrac{5252H}{R}$.

> T = torque, H = horsepower, and R = engine revolutions per minute (RPM).
>
> Find the RPM's for a 1200 horsepower engine with 870 foot pounds of torque, rounded to the nearest whole number.

20. The reactance offered by a capacitor in electronics is $X = \dfrac{1}{2\pi fC}$.

> X = reactance measured in ohms, f = frequency measured in cycles per second (hertz), C = capacitor size measured in farads.
>
> Find the capacitor size necessary for a circuit with a frequency of 120 hertz and a reactance of 18.37 ohms, rounded to six decimal places.

21. The formula to calculate inductive reactance is: $X = 2\pi fL$
    X = inductive reactance measured in ohms
    f = frequency measure in hertz (Hz)
    L = inductance measured in henrys (H)

    Calculate the inductance for an AC circuit with a frequency of 4.2 kHz and an inductive reactance of 457 Ω. Note: kHz has a metric prefix that must be considered.
    Round to 3 decimal places.

2.2: Solving formulas for different variables

In this section we will do math primarily with letters (variables). This can cause you to wonder why you signed up for this course, but rightly understood, this section is significantly easier than section 2.2 since there is no such thing as math with letters. b x c does not require us to do anything until we replace the variables with numbers. Although it can be intimidating to stare at an equation made entirely of letters and be asked to do something, there is really nothing you can do except move things around. For example, solve the area of a circle formula $A = \pi r^2$ for r.

$A = \pi r^2$

$\frac{A}{\pi} = r^2$ divide both sides by π

$r = \sqrt{\frac{A}{\pi}}$ take the square root of both sides

This new formula would be useful if we knew the area of a circle that we wanted but needed to calculate the necessary radius to produce it.

Consider the equation for the volume of a sphere:

Example 2.2.1: Volume of a sphere

Solve the volume of the sphere formula $V = \frac{4}{3}\pi r^3$, for r.

Solution:

$V = \frac{4\pi r^3}{3}$

$3V = 4\pi r^3$ multiply both sides by 3

$\frac{3V}{4\pi} = r^3$ divide both sides by 4π (since they are both

               multiplied by r they can be removed at the same

               time)

$r = \sqrt[3]{\frac{3V}{4\pi}}$ take the cube root of both sides to undo the cube

Final Answer: $r = \sqrt[3]{\frac{3V}{4\pi}}$

Note: This would be a useful formula for a welder wanting to produce spherical water tanks with particular volumes.

Example 2.2.2: Surface area of a cylinder

Solve the surface area of a cylinder formula $S = 2\pi r^2 + 2\pi rh$, for h.

Solution:

$S = 2\pi r^2 + 2\pi rh$

$S - 2\pi r^2 = 2\pi rh$ the entire expression $2\pi r^2$ is added to $2\pi rh$, and can be subtracted from both sides

$\frac{S - 2\pi r^2}{2\pi r} = h$ $2\pi r$ is multiplied by h, and can be divided from both sides

Final Answer: $h = \frac{S - 2\pi r^2}{2\pi r}$.

Note: Although this is a complicated formula, since h is wrapped with a big box and a lot of paper, it is simple to unwrap because there is no math to do. Note that it would be a lot more trouble to solve for r since it is in the formula twice and one is squared. Advanced algebra courses will give you the skill to take on this notable challenge.

There is added complexity when asked to solve for a variable in the denominator:

Example 2.2.3: Engineering a footing size

Solve the footing formula from section 1.5, $S = 12\sqrt{\frac{W}{B}}$, for B.

Solution:

$S = 12\sqrt{\frac{W}{B}}$

$\frac{S}{12} = \sqrt{\frac{W}{B}}$ start by dividing by 12 since multiplication is before exponents when unwrapping (remember that a square root is an exponent of $\frac{1}{2}$)

$\left(\frac{S}{12}\right)^2 = \frac{W}{B}$ square both sides to remove the square root

$B\left(\frac{S}{12}\right)^2 = W$ We need to get B out of the denominator since we want to solve for B not $\frac{1}{B}$. Multiplying both sides by B would bring it to the numerator (top).

$B = \frac{W}{\left(\frac{S}{12}\right)^2}$ since B is multiplied by the expression $\left(\frac{S}{12}\right)^2$, we just need to divide to remove it

Although correct, this equation is clumsy since it has fractions divided by fractions. It can be "cleaned up" if you know how to divide fractions. $B = W \div \left(\frac{S}{12}\right)^2 = W \div \frac{S^2}{12^2} = W \times \frac{12^2}{S^2} = \frac{144W}{S^2}$.

Final Answer: $B = \frac{144W}{S^2}$

Section 2.2: Solving Formulas for Different Variables

1. The <u>Ohm's Law and Watt's Power Formula Wheel of Equations</u> expresses the relationship between Power, Voltage, Current, and Resistance for electrical calculations.

   Two basic formulas in the table: **Ohm's Law V = R • I** and **Watt's Power Formula P = V • I** are the building blocks for the other 10. Use these two formulas to create the other 10 shown within the Wheel of Equations.

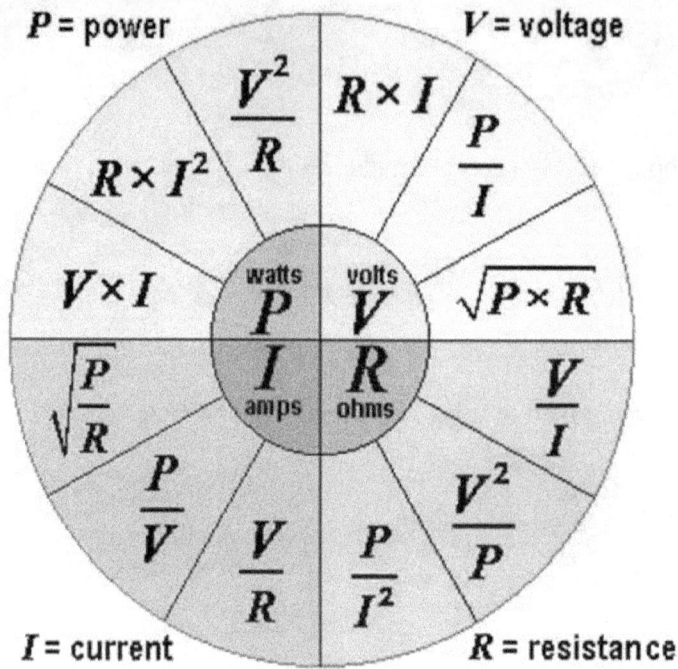

2. Solve the Beer's Law absorbance formula A = EcL, for E

3. Solve the automotive cornering force equation $F = \dfrac{v^2}{32r}$, for v.

4. Solve the automotive cornering force equation $F = \dfrac{v^2}{32r}$, for r.

5. Solve the octagonal window area formula in construction A = 4.828n^2, for n.

6. Solve the volume of cylinder formula V = $\pi r^2 h$, for h.

7. Solve the temperature formula F = 1.8C + 32, for C

8. Solve the volume of cylinder formula V = $\pi r^2 h$, for r.

9. Solve the allowable stress on a post formula in construction $S = \dfrac{3ED^2}{10L^2}$, for D.

10. Solve the allowable stress on a post formula in construction $S = \dfrac{3ED^2}{10L^2}$, for L.

11. Solve the moment formula for a simple beam $M = \dfrac{wL^2}{8}$, for L.

12. Solve the shear formula for a simple beam $V = w\left(\dfrac{L}{2} - x\right)$, for L.

13. Solve the engine displacement formula $D = \dfrac{\pi b^2 sc}{4}$, for s.

14. Solve the engine displacement formula $D = \dfrac{\pi b^2 sc}{4}$, for b.

15. Solve the horsepower formula $H = W\left(\dfrac{S}{234}\right)^3$, for S.

16. Solve the formula for the speed of a car $S = \dfrac{DR}{336G}$, for D.

17. Solve the formula for the speed of a car $S = \dfrac{DR}{336G}$, for G.

18. Solve the piston speed formula $P = \dfrac{SR}{6}$, for R.

19. Solve the carburetor size formula $C = \dfrac{dRV}{3456}$, for V.

20. Solve the torque formula $T = \dfrac{5252H}{R}$, for H.

21. Solve the torque formula $T = \dfrac{5252H}{R}$, for R.

22. **Challenge Problem:** Solve the formula for resistance in parallel electrical circuits
$\dfrac{1}{R_t} = \dfrac{1}{R_1} + \dfrac{1}{R_2}$, for R_t.

2.3: Solving complex equations

In this section we will include equations with exponents and parentheses. You now need to understand the inverse operation for square (^{2}) is square root ($\sqrt{}$), cube (^{3}) is cube root ($\sqrt[3]{}$) and fourth (^{4}) is fourth root ($\sqrt[4]{}$). You also need to be able find and use these operations on your calculator. Before we get into some examples, recall from section 2.1 the order of operations and that removal must occur in the reverse order:

1. **P**arentheses
2. **E**xponents
3. **M**ultiply/**D**ivide
4. **A**dd/**S**ubtract

The goal of this section is to widen our scope to include any equation we might encounter.

Consider the following example from chapter 1 involving the cornering force a vehicle exerts on its passengers as a function of the radius of the turn and speed of the vehicle:

Example 2.3.1: Cornering force for a vehicle

Solve for velocity (v) measured in feet per second if you know the force (F) = 2 g's and the radius(r) = 4 feet. Use the formula: $F = \dfrac{v^2}{32r}$.

Solution:

$2 = \dfrac{v^2}{32*4}$ substitute the numbers into the formula

$2 = \dfrac{v^2}{128}$ simplify

$256 = v^2$ multiply both sides by 128

$16 = v$ the inverse operation for square is square root ($\sqrt{}$ on your calculator)

Final Answer: The velocity (v) = 16.

Consider a very practical and complex example of money growing at a monthly interest rate:

Example 2.3.2: Calculating compound interest rates

Solve for the interest rate (r), if the amount (A) = $1188, principal (P) = $1080, and the time (M) = 4 months. Use the formula: $A = P(1 + \frac{r}{12})^M$.

Solution:

$1188 = 1080(1 + \frac{r}{12})^4$	substitute the numbers into the formula
$1.1 = (1 + \frac{r}{12})^4$	The parenthesis must be saved for last. Outside it are multiplication and an exponent, so the multiplication is removed first by dividing both sides by 1080.
$1.024 \approx 1 + \frac{r}{12}$	Remove the exponent (^{4}) by taking the fourth root ($\sqrt[4]{}$) of both sides. This can be done on your calculator by entering 4 ($\sqrt[x]{}$) 1.1.
$.024 \approx \frac{r}{12}$	now we can work on removing the numbers inside the parenthesis, addition is removed first
$r \approx .288$	multiply both sides by 12

Final Answer: Changing to a percent, r ≈ 28.8%.

Note: You would have to get a 28.8% interest rate for $1080 to grow to $1188 in 4 months.

Consider another equation used to calculate roofing for a house, involving a large square root sign:

Example 2.3.3: Calculating the slope of a roof

Solve for the slope of the roof (s) if the square feet of roofing (R) = 2400, area (A) = 2000. Use the formula: $R = A\sqrt{1 + s^2}$.

Solution:

$2400 = 2000\sqrt{1 + s^2}$	substitute the numbers into the formula
$1.2 = \sqrt{1 + s^2}$	This problem is tricky in that it is tempting to remove the 1 first, but the large square root symbol acts as a parenthesis and so removing the 1 and the (^{2}) must be saved for last. Start by removing 2000, divide both sides by 2000.
$1.44 = 1 + s^2$	square both sides to remove the square root
$.44 = s^2$	once inside the parenthesis, remove the 1 by subtracting 1 from both sides
$s \approx .663$	take the square root of both sides

Final Answer: The slope of the roof is approximately .663.

Note: Roof slopes are fractions with 12 for a denominator. Solve the equation $.663 = \frac{x}{12}$, for x. The decimal .663 is very close to 8/12, which is a common roof slope.

Consider an example where there are two representations of the letter (variable) that must be solved for:

Example 2.3.4: Costing for a welding job

A welder who has $1460 of overhead costs per month can build railings at a production cost of $120 each. The selling price per railing as a function of the number of railings produced is modeled by:
$$P = \frac{1460+120N}{N}.$$
Find the number (N) of railings that he must build to charge a price (P) of $197 each to cover his production costs and overhead.

Solution:

$197 = \frac{1460+120N}{N}$	substitute the numbers into the formula
197N = 1460 + 120N	To solve for N it is necessary to collect the N's together, which is done by adding or subtracting them. The justification for this is the distributive property. Begin by multiplying both sides of the equation by N.
77N = 1460	subtract 120N from both sides
N ≈ 19	divide both sides by 77

Final Answer: Producing and selling 19 railings for a price of $197 will cover production costs and overhead. Knowing break-even production and pricing keeps a company from losing money and provides a guideline for increasing production or price to make a profit.

Section 2.3: Solving Complex Equations

1. In electronics, Power (P) is $P = I^2 R$.

   P = power measured in watts, R = resistance measured in ohms, and I = current measured in amps.

   Find the current passing through a circuit that consumes 120 watts of power with a resistance of 10 ohms, rounded to one decimal place.

2. The moment for a simple beam, uniformly loaded, can be found using the formula: $M = \dfrac{wL^2}{8}$.

   Find length (L) in feet if the moment (M) = 894 ft^2-lbs and weight (w) = 87 pounds, rounded to one decimal place.

3. A manufacturer's sales price per part (P) can be calculated by: $P = \dfrac{350 + 14.25N}{N}$.

   Where P = minimum sales price per part, $350 is the overhead costs, $14.25 is the cost of producing the part, and N = number of parts produced.

   Find the number of parts that must be produced to make the sales price $20, rounded to the nearest part.

4. Spacing for fence slats (S) can modeled by: $S = \dfrac{W - BN}{N+1}$.

   S = spacing between slats, W = width between posts, N = number of slats, and B = width of one slat.

   Find the number of slats that will fit if W = 95 inches, B = $3\frac{1}{2}$ inches & S = $\frac{1}{2}$", rounded to the nearest slat.

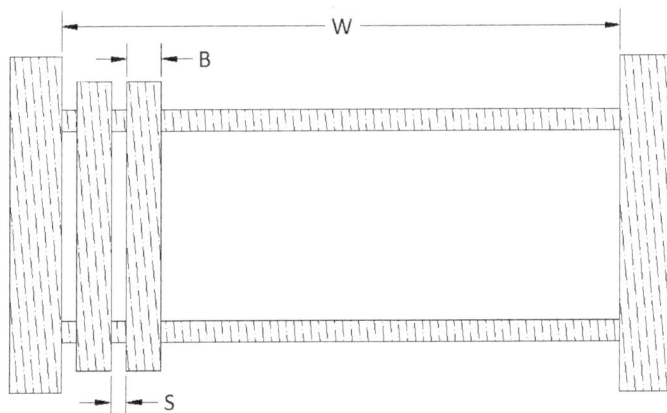

5.   The formula for engine displacement is $D = \dfrac{\pi b^2 sc}{4}$.

   D = engine displacement measured in cubic inches, b = bore (diameter of the cylinder) measured in inches, s = stroke (distance that the piston travels) measured in inches, and c = number of cylinders.

   Calculate the bore for an 8-cylinder engine with 370 cubic inches of displacement and a 5-inch stroke, rounded to three decimal places.

6.   The allowable stress (S) on a post is $S = \dfrac{3ED^2}{10L^2}$.

   S = allowable stress measured in pounds per square inch (PSI), D = dimension of the post measured in inches, L = length of the post measured in inches, and E = elasticity of the beam measured in pounds per square inch (PSI).

   Find the dimension of a post rounded up to the nearest whole number, if S = 420, E = 1,500,000, L = 96.

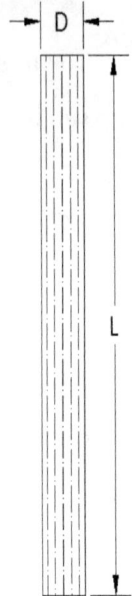

7.   The formula for horsepower is $H = W\left(\dfrac{S}{234}\right)^3$.

   H = horsepower, W = weight, and S = speed.

   Calculate the speed a car will be capable of that weighs 2460 pounds with 390 horsepower, rounded to the nearest mile per hour (MPH).

8.   The moment of inertia (I) of a beam is $I = \dfrac{bd^3}{12}$.

   Note: Moment of inertia is a measure of a beam's effectiveness at resisting bending based on its cross-sectional shape.

   I = moment of inertia of the beam measured in inches^4, b = width of the beam measured in inches and d = height of the beam measured in inches.

   Find the height of a beam rounded to the nearest 8th of an inch if b = $7\dfrac{1"}{4}$ and I = 6.5.

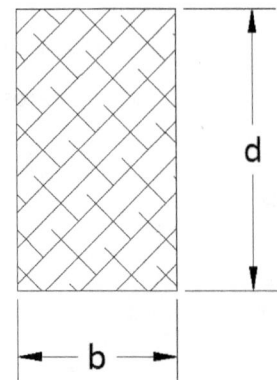

9. The point load deflection (D) of a beam is $D = \dfrac{PL^3}{48EI}$.

Note: Deflection is simply a measurement of the amount of bend in a beam.

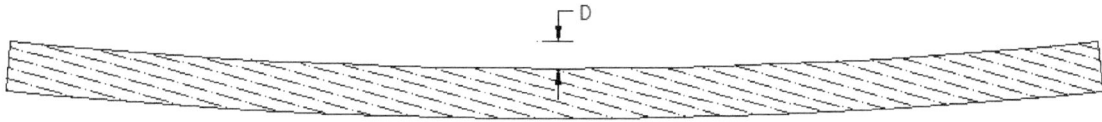

D = deflection measured in inches, P = weight on the beam measured in pounds, L = length of the beam measured in inches, E = elasticity of the beam measured in pounds per square inch (PSI), and I = moment of inertia of the beam measured in inches^4.

Find the length of a beam rounded to the nearest inch if D = .9, P = 3800, E = 1,700,000, and I = 326.

10. The formula to calculate impedance in an RL circuit is: $Z = \sqrt{R^2 + X^2}$.
    Z = impedance measured in ohms
    R = resistance measured in ohms
    X = reactance measured in ohms

Calculate the resistance in a circuit with 6.2 kΩ of impedance and a reactance of 4.1 kΩ, rounded to the nearest tenth of a kΩ.

11. The formula to calculate size (S in inches) of a square footing is $S = 12\sqrt{\dfrac{W}{B}}$.

W = weight on the footing (in pounds) & B = soil bearing capacity in pounds per square foot (PSF).

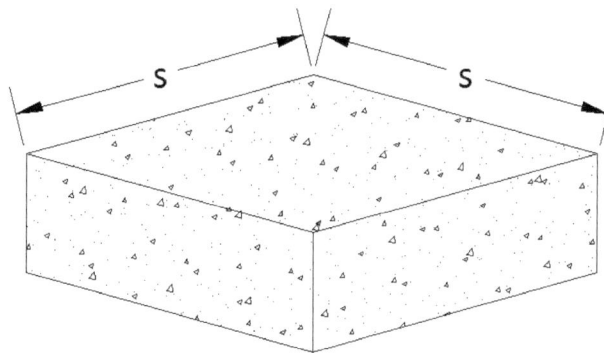

Find the weight that a 22" by 22" footing can support that sits on soil able to bear 1200 psf. Round your answer to the nearest pound.

12. Fill in the table of values accurate to two decimal places for the electrical circuit wired in **series**, using the two primary electrical formulas:

Ohm's Law V = R • I   and   **Watt's Power Formula P = V • I**

V = voltage (volts), I = current (amps), R = resistance (ohms), P = power (watts)

What you need to know about series circuits:

a. Electricity must pass through both resistors.

b. Ω is the symbol for ohm, which is the unit of measurement for resistance R.

c. The subscripts for the letters serve only to distinguish to which resistor they belong: R_1 is resistor one.

d. $R_1 + R_2 = R_{total}$

e. $V_1 + V_2 = V_{total}$

f. $I_1 = I_2 = I_{total}$

g. $P_1 + P_2 = P_{total}$

$R_1 .48 \, \Omega$

12 V

$R_2 .56 \, \Omega$

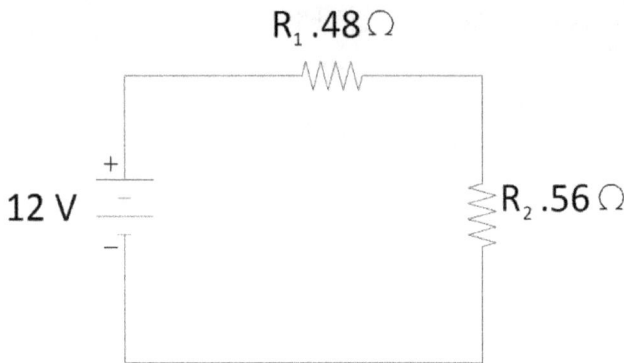

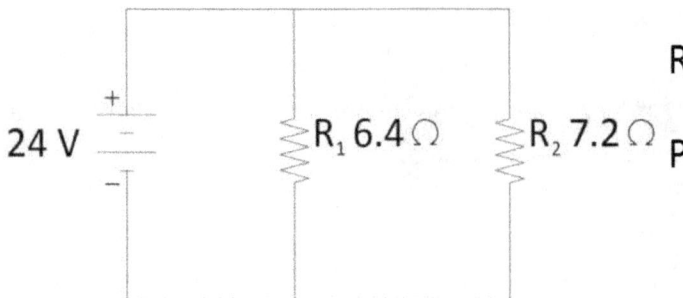

	Total	R_1	R_2
V			
I			
R			
P			

13. Fill in the table of values accurate to three decimal places for the electrical circuit wired in **parallel**, using the two primary electrical formulas:

Ohm's Law V = R • I   and   **Watt's Power Formula P = V • I**

V = voltage (volts), I = current (amps), R = resistance (ohms), P = power (watts)

What you need to know about parallel circuits:

a. Electricity passes through one or the other resistor.

b. Ω is the symbol for ohm, which is the unit of measurement for resistance R.

c. The subscripts for the letters serve only to distinguish to which resistor they belong: R_1 is resistor one.

d. $\dfrac{R_1 R_2}{R_1 + R_2} = R_{total}$

e. $V_1 = V_2 = V_{total}$

f. $I_1 + I_2 = I_{total}$

g. $P_1 + P_2 = P_{total}$

24 V

$R_1 \, 6.4 \, \Omega$ $R_2 \, 7.2 \, \Omega$

	Total	R_1	R_2
V			
I			
R			
P			

14. **Challenge Problem:** Fill in the table of values accurate to two decimal places for the electrical circuit wired in **series/parallel**, using the two primary electrical formulas:

    Ohm's Law V = R • I and **Watt's Power Formula P = V • I**

    V = voltage (volts), I = current (amps), R = resistance (ohms), P = power (watts)

    What you need to know about series/parallel circuits:

    a. $R_1 + \dfrac{R_2 R_3}{R_2 + R_3} = R_{total}$

    b. $V_2 = V_3$ and $V_1 + V_2 = V_{total}$

    c. $I_1 = I_{total}$ and $I_2 + I_3 = I_{total}$

    d. $P_1 + P_2 + P_3 = P_{total}$

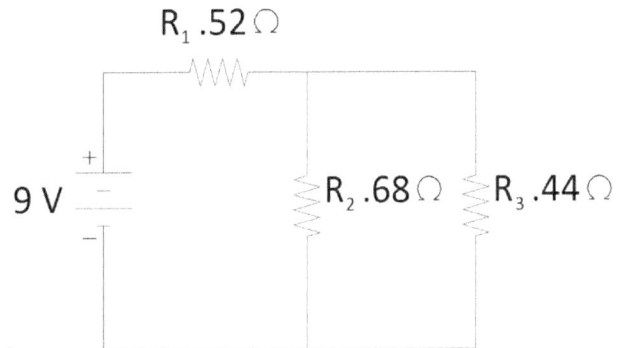

R_1 .52 Ω

9 V

R_2 .68 Ω R_3 .44 Ω

	Total	R_1	R_2	R_3
V				
I				
R				
P				

Chapter 3:
Right Triangle Geometry

In chapters 1 and 2 we learned how algebra can be used to figure out any unknown quantity in a formula. In this chapter we will narrow our focus to formulas that pertain to right triangles. Right triangles appear in a surprising number of welding, electronics, diesel, construction, automotive, and manufacturing applications.

3.1: Angles

We now turn our attention to angles, in preparation for trigonometry in section 3.3. Angles are commonly measured in degrees using a protractor. History provides an interesting answer to the questions, what is a degree and why is a circle composed of 360 of them? Ancient astronomers are responsible for dividing the circle into 360°, due to the fact that their calendar had 12 months of 30 days each resulting in a 360-day year. A degree was used as a measure for the angle the earth traveled in one day in its circular path around the sun.

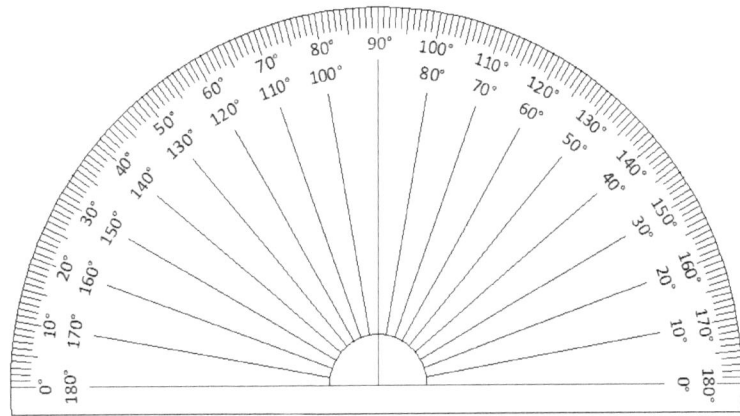

The protractor can be aligned with 0° on either ray (side) of the angle. If the 0° mark on the left of the protractor is used then you will be reading the outer set of numbers. If the 0° mark on the right of the protractor is used then you will be reading the inner set of numbers.

If you are measuring the angle below, you can align the protractor in either way illustrated to obtain 73°.

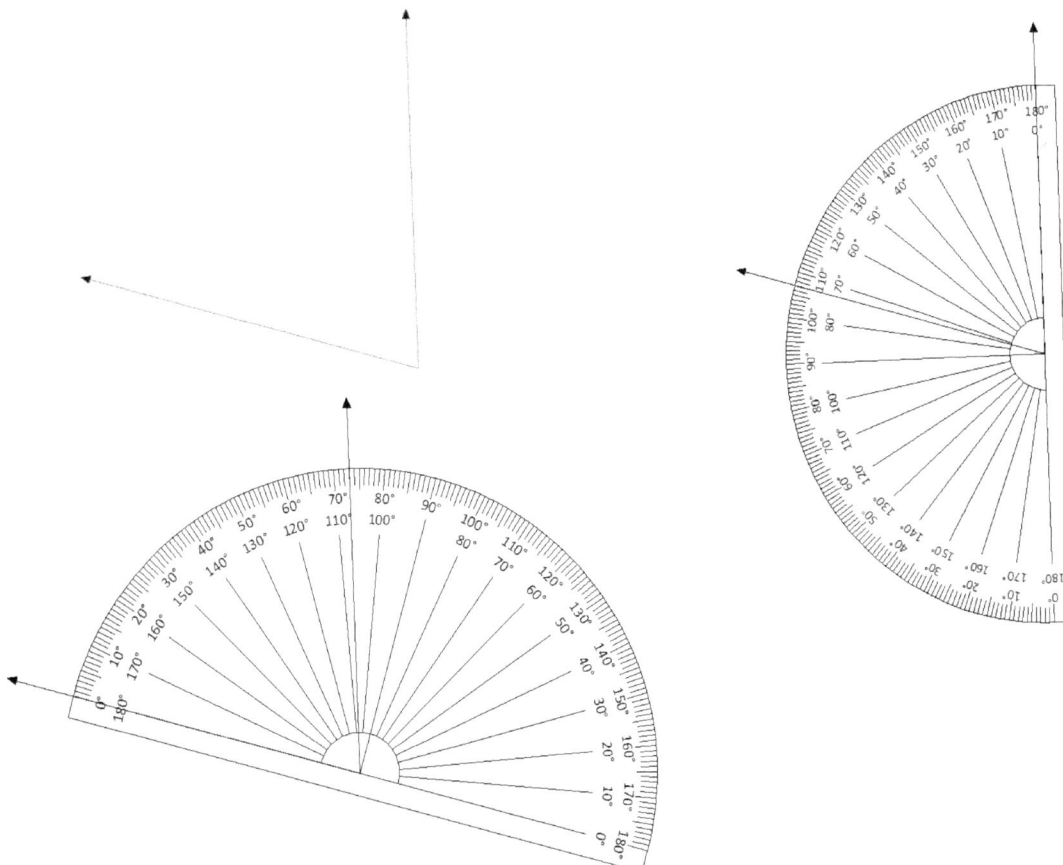

There are only two requirements for measuring an angle:

1. The center of the protractor has to be on the vertex (or point) of the angle.
2. One of the rays is aligned with either $0°$ mark.

> **Angle Facts:**
>
> 1. **Turning a complete circle is 360°**
> 2. **The angles in a triangle add up to 180°**
> 3. **A straight line is 180°**
> 4. **Parallel lines have equal angles**

The last fact may require a picture. Notice the parallel lines create two groups of angles that are all the same.

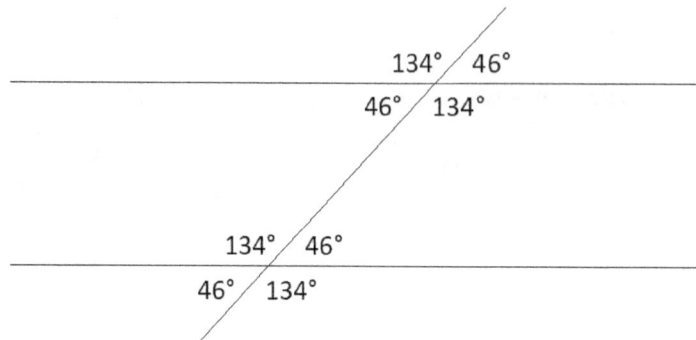

$$
\begin{array}{cc}
134° & 46° \\
46° & 134°
\end{array}
$$

$$
\begin{array}{cc}
134° & 46° \\
46° & 134°
\end{array}
$$

Since a straight line is $180°$, notice the two angles add to $180°$ as well. Therefore, it is only necessary to know one of the eight angles in order to find the others. This simple and elegant fact can be difficult to apply and all the subtle relationships here are overwhelming when put into words. Study the figure carefully.

Apply the angle facts to solve the following practical problems:

Example 3.2.1: Calculating roof angles

To make the necessary cuts to build the roof structure the angles labeled A through F must be calculated. The angle of the roof is 32°.

Lines that appear parallel really are. The supporting posts are rectangular. Since the left and right halves are symmetrical, corresponding angles that are mirror images will be equal.

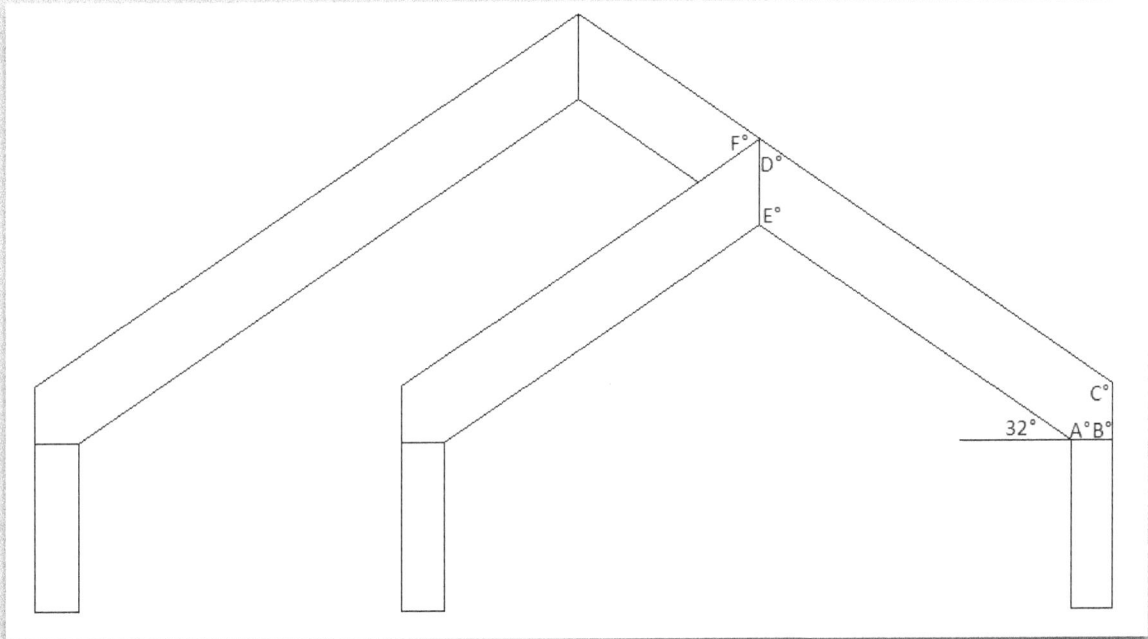

Solution:

A = 148°   32° + A = 180° (angle fact 3)

B = 90°   from angle fact 3 and the rectangular post with 90° angles

It is often necessary to get creative calculating angles. Adding the dashed lines to the figure creates a right triangle that will unlock the mystery.

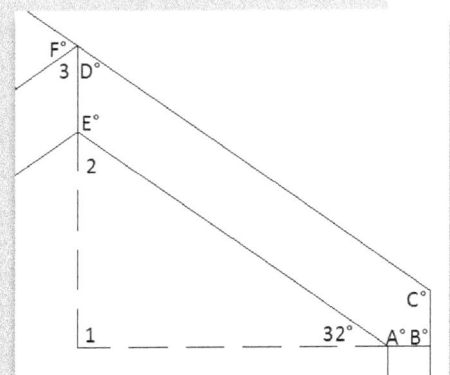

angle 2 = 58°   1 is 90°, so 32° + 90° + 2 = 180°
                  (angle fact 2)

E = 122°   2 + E = 180° (angle fact 3)

D = 58°   angle 2 = D° (angle fact 4)

C = 122°   Next D + C = 180° (angle fact 4)

F = 64°   3 = 58° because of the symmetry of angles 3 and D,
                  and D + 3 + F = 180° (angle fact 3)

Final Answers: A = 148°, B = 90°, C = 122°, D = 58°, E = 122°, F = 64°

Note: F is a very important and challenging angle to calculate for carpenters.

Example 3.2.2: Calculating angles for CNC (computer numerically controlled)

The symmetrical sheet metal sign can only be produced if the angles are known. Calculate angles A through D.

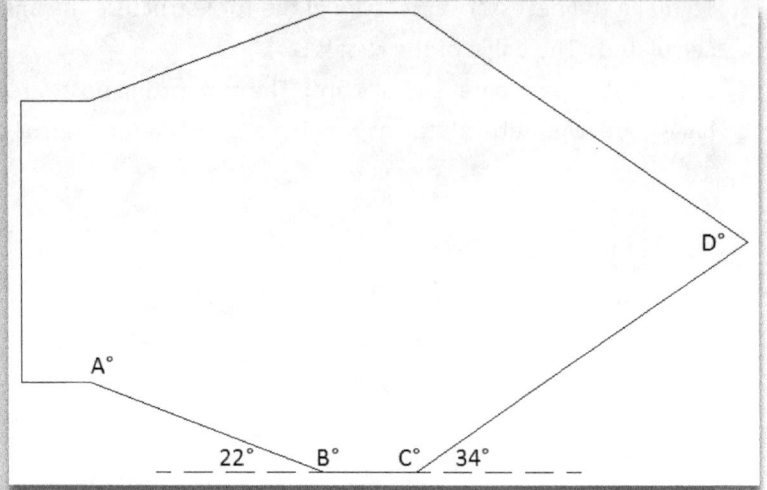

Solution:

$B = 158°$ $22° + B = 180°$ (angle fact 3)

$C = 146°$ $34° + C = 180°$ (angle fact 3)

Again it is necessary to draw in some extra lines to make further progress.

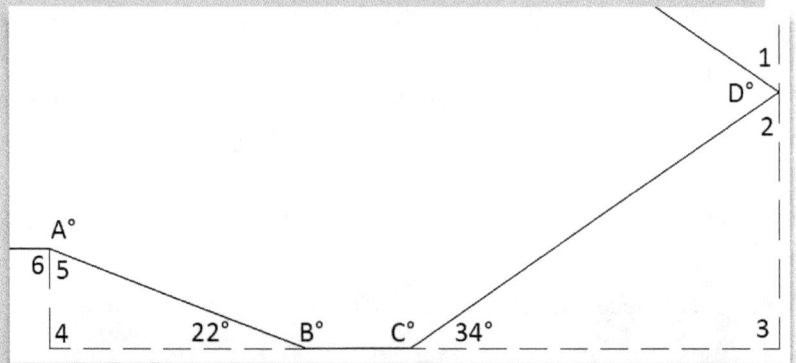

angle 2 = 56° $3 = 90°$ and $2 + 3 + 34° = 180°$ (angle fact 2)

$D = 68°$ $1 = 2$ based on symmetry and $1 + D + 2 = 180°$ (angle fact 3)

Angle 5 = 68° $4 = 90°$ and $5 + 4 + 22° = 180°$ (angle fact 2)

$A = 202°$ $6 = 90°$ and $5 + 6 + A = 360°$ (angle fact 1)

Final Answers: $A = 202°, B = 158°, C = 146°, D = 68°$

Section 3.3.1 Angles

1. Measure angular dimensions A through J using a protractor on the shape to the nearest degree.

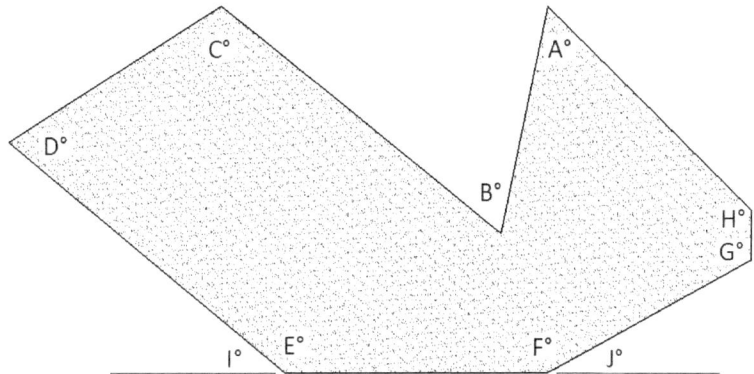

2. Calculate angles A through D on the direction arrow.

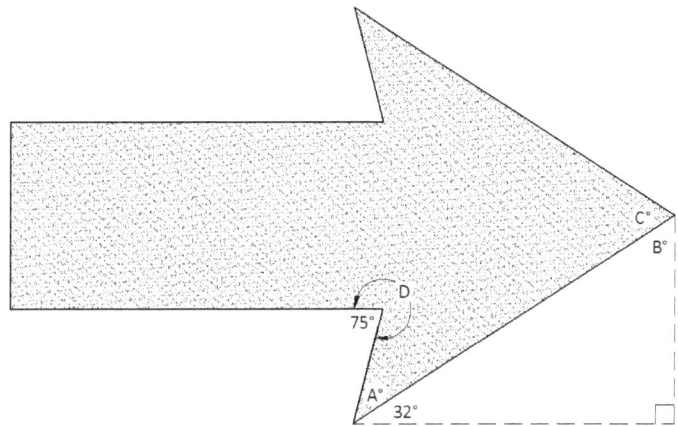

75°

D

32°

3. Calculate angles A, B and C on the truss.

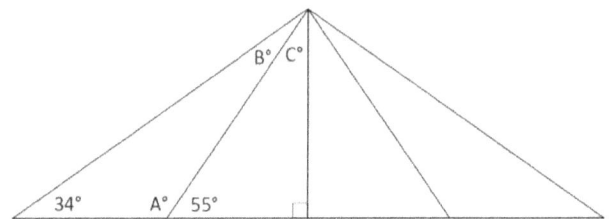

34°   A° 55°

B° C°

4. Calculate angles A through F in the roof of the house. Note: The rafters are parallelograms and so the opposite lines are parallel.

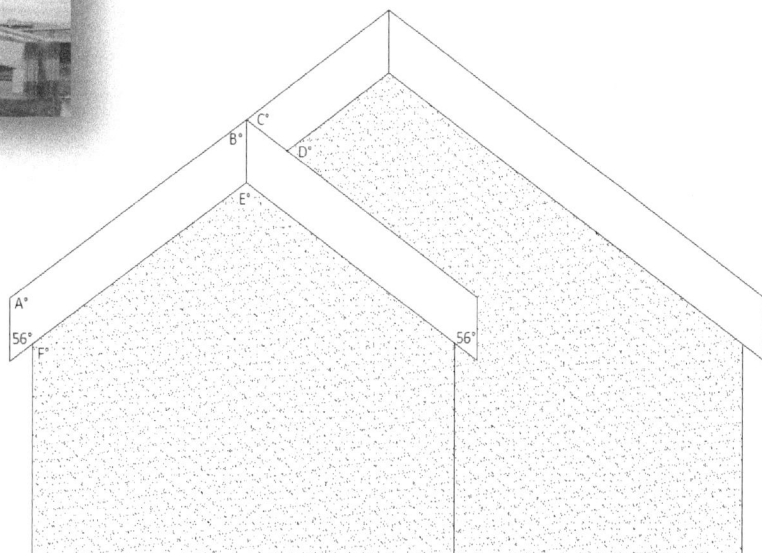

56°

56°

87

5. Calculate angle D if the holes are evenly spaced around the circle.

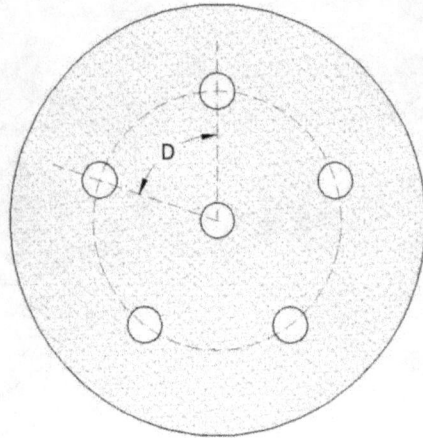

6. Calculate angles A-E in the rafter. Note: The dashed lines are parallel.

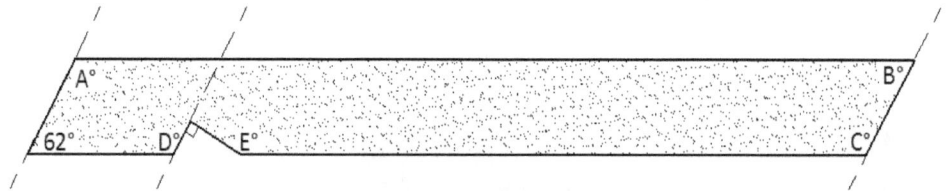

7. Fifteen holes are to be evenly spaced in a circular pattern in a gear in a counter clockwise direction from a starting line. Find the angle to each hole as measured from the starting line.

Hole #	Angle
1	
2	
3	
4	
5	
6	
7	
8	
9	
10	
11	
12	
13	
14	
15	

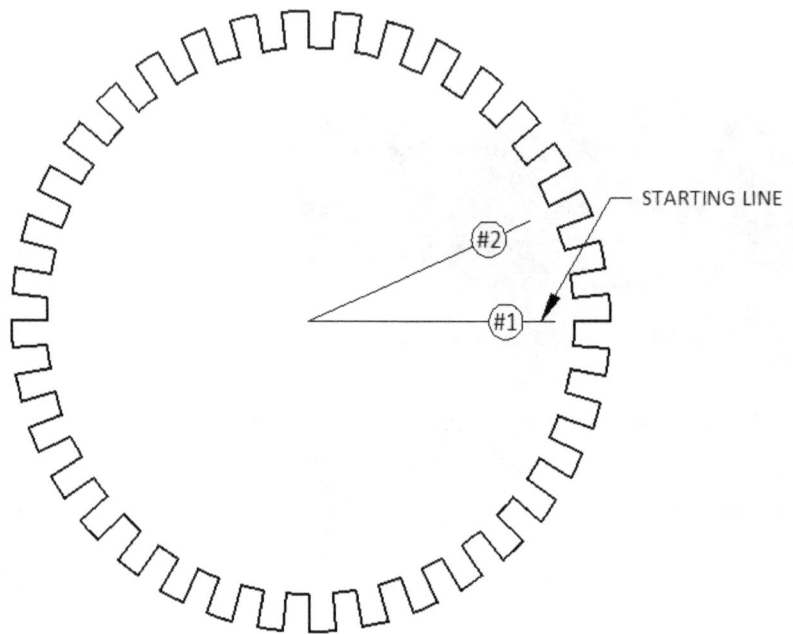

STARTING LINE

#2

#1

3.2: Perimeter and Area

Skill in calculating the area of rectangles, triangles, circles, and trapezoids is essential for ordering materials and bidding jobs in every technical trade. A drywall contractor must take measurements of rectangular ceilings and windows, triangular, and trapezoidal walls, and be able to translate these measurements into areas for material orders and bids. Perimeter is simply the distance around something (often measured in lineal feet). Perimeter is one-dimensional because it is a measure of length (feet, meters, or inches). Determining the amount of metal railing for a deck, lane striping for a track or trim around a window, are all questions of perimeter. Area is two-dimensional because it is the measure of a region (square feet, square meters, or square inches). Determining the amount of paint for a wall, carpet for a room or grass for landscaping are all questions of area.

Consider the following practical example of a rectangular roof:

Example 3.2.1: Perimeter and area for the rectangular roof

1. Calculate the lineal footage of fascia necessary to surround the shed roof.
2. Calculate the number of 4' x 8' sheets of plywood necessary to cover the shed roof.

Solution:

1. There is a formula in the appendix for the perimeter of a rectangle, but …. do we need it?
   38 ft + 38 ft + 74 ft + 74 ft = 224 ft.

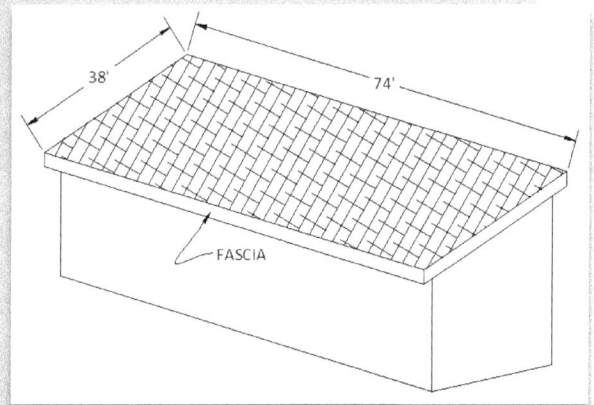

2. The formula for the area of a rectangle is $A = LW$.
   74 ft x 38 ft = 2812 ft^2 this is the area of the roof
   2812 ft^2 $\times \frac{1\ \text{sheet}}{32\ \text{ft}^2}$ = 87.875 sheets   the area of one sheet of plywood is 4 ft x 8 ft = 32 ft^2

Final Answers: 224 lineal feet of fascia, and 88 sheets of plywood

> Note: in this case, it would make sense to round the number of sheets up to the next whole number regardless of the decimal part of the answer. We would order 88 sheets even if the math had resulted in 87.14, for example. Though 87.14 is closer to 87 than to 88, the situation demands that we obtain more than enough plywood, rather than too little. Rounding is a dynamic skill where the situation determines the need for accuracy.

Example 3.2.2: Perimeter and area for a floor

1. Calculate the lineal footage of base molding necessary to surround the room.
2. Calculate the area of the floor and determine how many lineal feet of 3-inch wide hardwood are necessary to cover the floor.

Solution:

1. There are 10 lengths to add up, but a wonderful shortcut for rectangular shapes involves observing that since there are 25 ft across the "top" of the room there will be 25 ft across the "bottom". Also, there are 21 ft on the left and there will be 21 ft on the right. 2(25) + 2(21) = 92 ft. It is a good idea to add all 10 lengths to check this result.

2. There is no formula for the area of such a complex shape, but area is still easy to determine. Find the area of the 21 ft x 25 ft rectangle that surrounds the whole room, and then subtract the areas of the three smaller rectangles that are not part of the room.

   $(21 \times 25) - (8 \times 6) - (5 \times 12) - (5 \times 9)$

   $525 - 48 - 60 - 45 = 372 \text{ ft}^2$

   Based on the picture of 1 ft^2 of flooring it will take 4 lineal feet to cover 1 ft^2.

   $$\frac{372\,ft^2}{1} \times \frac{4\,ft}{1ft^2} = \frac{372\,\cancel{ft^2}}{1} \times \frac{4\,ft}{1\cancel{ft^2}} = 1488 \text{ ft.}$$

Final Answers: The floor will require 92 lineal feet of base molding, and 1488 lineal feet of 3" wide hardwood.

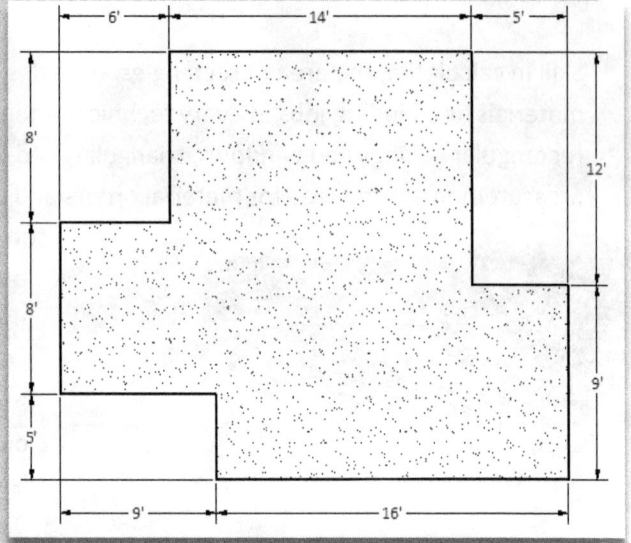

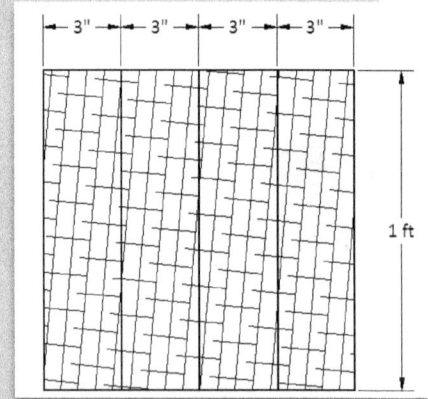

It takes more effort, but it is good exercise to divide the floor into rectangles; add up the areas and show that you still get 372 ft^2. This would represent only one possible method for dividing it up. Try another way and see that you get the same area. Most interesting areas involve breaking a complicated shape into the simpler shapes for which we have area formulas.

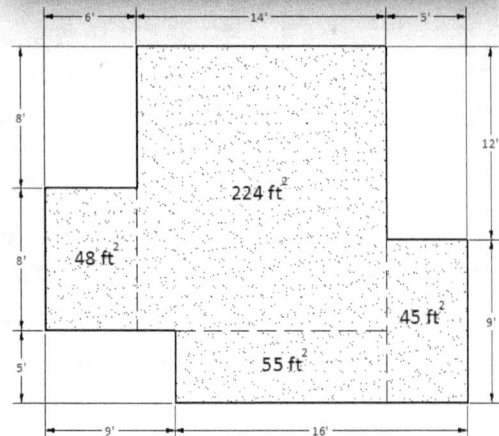

Example 3.2.3: Find the amount of siding for a wall

Calculate the number of pieces of Hardiplank siding necessary to cover the wall. Hardiplank is a common type of horizontal siding where each piece is 12 ft long and is spaced so the exposure of each piece is 7 inches.

drawing not to scale

Solution:

There is no formula for this shape but it divides easily into an 18 ft x 8 ft rectangle and the triangle up in the gable. The rectangle has area 144 ft^2. The formula for the area of a triangle is $A = \frac{1}{2}bh$ and the base is 18 ft. We need a proportion to find the height of the triangle:

$\frac{5}{12} = \frac{h}{9}$, h = 3.75 ft. Now that we know the height, $A = \frac{1}{2}$ x 18 x 3.75 = 33.75 ≈ 34 ft^2.

The total square footage for the wall is 144 + 34 = 178 ft^2.

Dimensional analysis can now help us change ft^2 to the number of pieces needed. The units would imply that we need to multiply by a fraction with pieces in the numerator (top) and ft^2 in the denominator (bottom).

7in x 144in = 1008 in^2 the area that one piece of siding will cover on the wall

$\frac{1008\,in^2}{1} \times \frac{1ft^2}{144in^2} = 7\,ft^2$ the area of one piece of siding measured in square feet

$\frac{178\,ft^2}{1} \times \frac{1\,piece}{7ft^2} = \frac{178\,ft^2}{1} \times \frac{1\,piece}{7ft^2} ≈ 25.4$ pieces.

Final Answer: 26 pieces

Note: It is remarkable how complicated a seemingly simple problem can be. You need to mull this over in your mind in order to deeply understand it. The 7-inch overlap of the siding made the problem significantly more difficult. If you cannot comfortably use the principles of dimensional analysis from section 1.4 this problem is likely to result in some broken pencils. You may need some review or extra practice to master this useful skill.

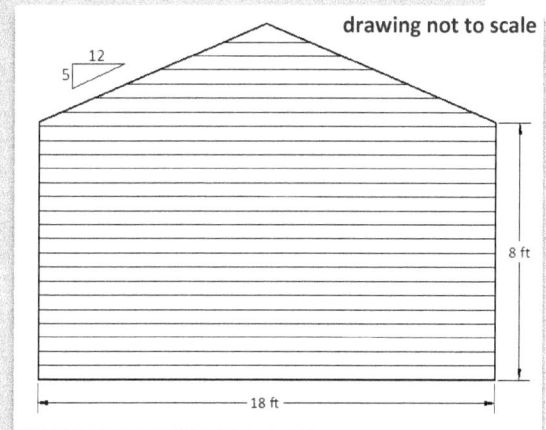

Side question: Now that we know the height of the triangle, use the Pythagorean Theorem to calculate the perimeter of the entire wall to be 53'-6"? A carpenter may need this calculation to place a material order for trim.

As noted in the introduction to this book, there is no shortcut if your goal is deeper than just getting through this class. Ownership comes at a cost. You should have some idea of how much you are owning from this course based on your reaction to the previous example.

Example 3.2.4: Circumfernce and area for a circular window

Calculate the area of the glass and the circumference of the outer edge of the trim around the circular window.

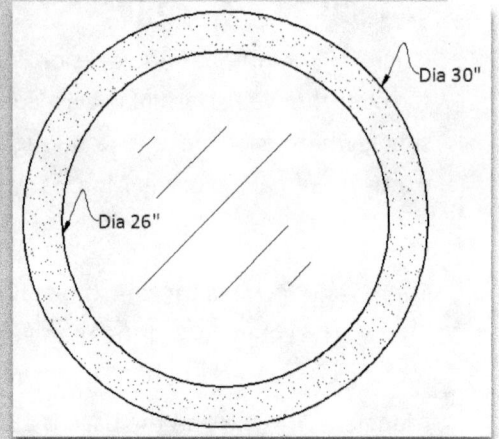

Solution:

$A = \pi r^2$ area of a circle formula

$A = \pi 13^2$ substitute the radius into the formula

$A = 169\pi \approx 530.9 \text{ in}^2 \approx 3.7 \text{ ft}^2$

$C = 2\pi r$ circumference of a circle formula (the perimeter of a curve is renamed circumference to distinguish it from the measurement of a straight line)

$C = 2\pi 15$ substitute the radius into the formula

$C = 30 \pi \approx 94.2 \text{ in} \approx 7.9 \text{ ft}$

Final Answers: The area of the glass is approximately 530.9 in² or 3.7 ft². The Circumference of the window is approximately 94.2 in. or 7.9 ft.

Example 3.2.5: Weight of a welded trailer

Find an approximation of the weight of the welded steel trailer using only the dimensions given, rounded to the nearest kilogram. Note: The type of steel is labeled on the diagram which can be interpreted using the steel design table in the appendix.

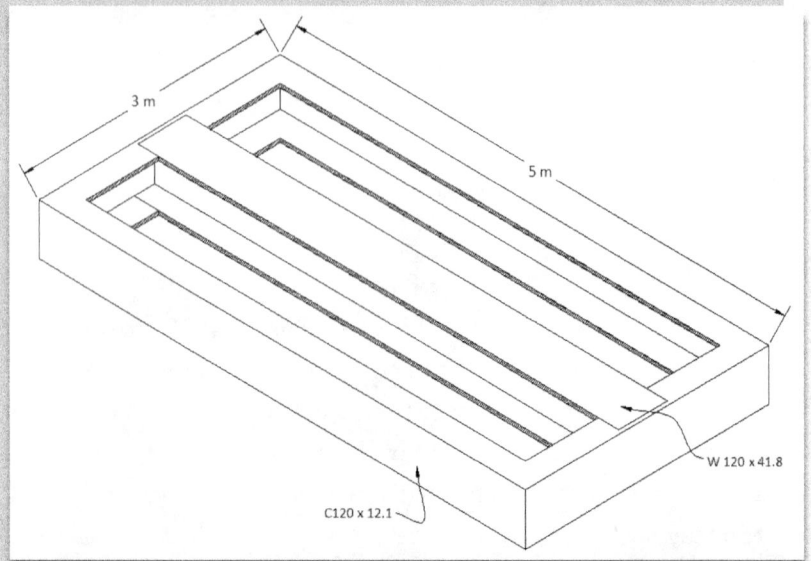

Solution:

Channel around the perimeter:

$16 \text{ m} \times \frac{12.1 \text{ kg}}{\text{m}} \approx 194 \text{ kilograms}$ there are 16 meters of channel around the perimeter of the trailer

I-beam through the center:

$5 \text{ m} \times \frac{41.8 \text{ kg}}{\text{m}} = 209 \text{ kilograms}$ there are 5 meters of wide flange I-beam

Final Answer: The weight of the trailer is approximately 403 kilograms.

Section 3.2: Perimeter and Area

1. Find the surface area of the wall without the window. A painter might need to perform such a calculation to order paint or price a job. Answer in square feet.

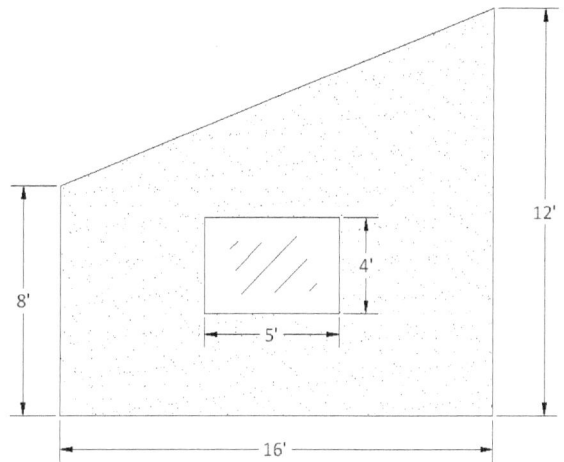

2. Find the number of 4' x 8' sheets of plywood that should be ordered for the roof, rounded up to the nearest whole sheet.

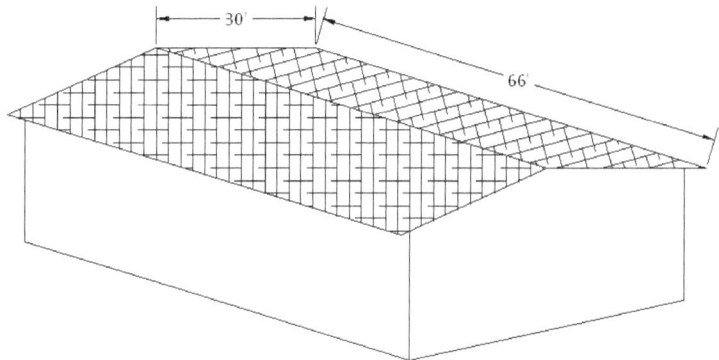

3. Find the number of 4" x 4" tiles that are needed for the kitchen floor (the shaded region).

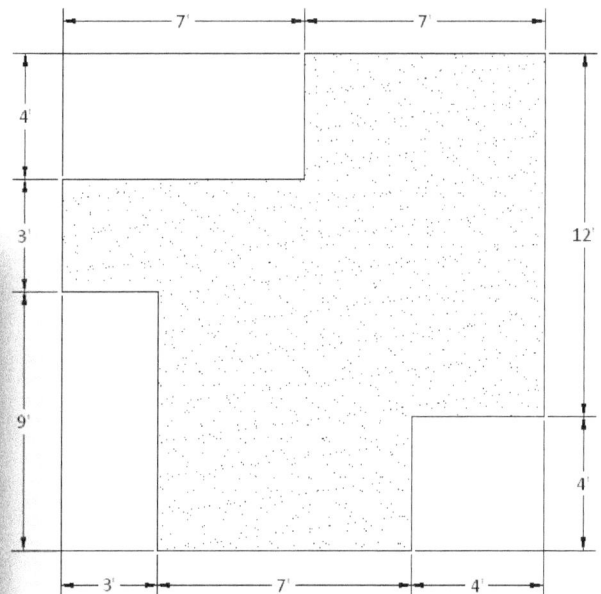

4. How many lineal feet of two inch wide hardwood flooring are necessary to cover the kitchen (the shaded region)?

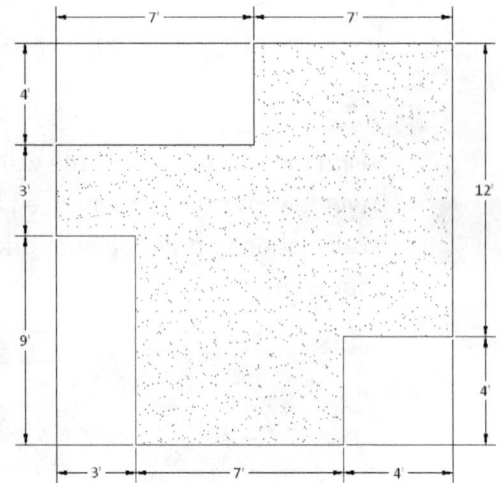

5. A standard all-weather track has eight lanes that are one meter wide, with 100-meter straightaways, two semi-circular ends, and a grass infield. Calculate the area of grass and calculate the surface area of the all-weather track. Round your answers to the nearest whole number.

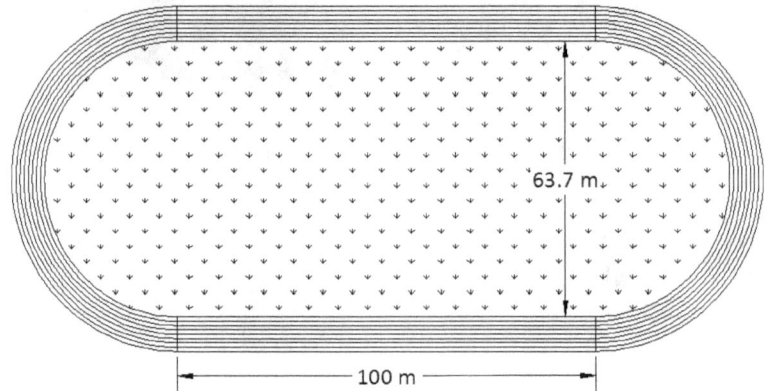

6. The circular top of a concrete test cylinder has a 6-inch diameter. If the top of the cylinder is able to withstand 72,400 pounds of force, calculate the psi strength of the concrete (recall that psi refers to pounds per square inch). Hint: You are only considering the weight and the area of the circle at the top of the cylinder. Round your answer to the nearest whole number.

7. Roofing for a house is ordered by the square foot. The formula for calculating the number of square feet of roofing for a house is: $R = A\sqrt{1 + S^2}$.

> Where R = number of square feet, A = area or square footage of the rectangular <u>floor</u> of the house, and S = the slope of the roof. Calculate the number of square feet of roofing for the house below with a roof slope of $\frac{10}{12}$, rounded to the nearest whole number. Hint: Hip and ridge lines are drawn in for effect but should be ignored for the calculation.

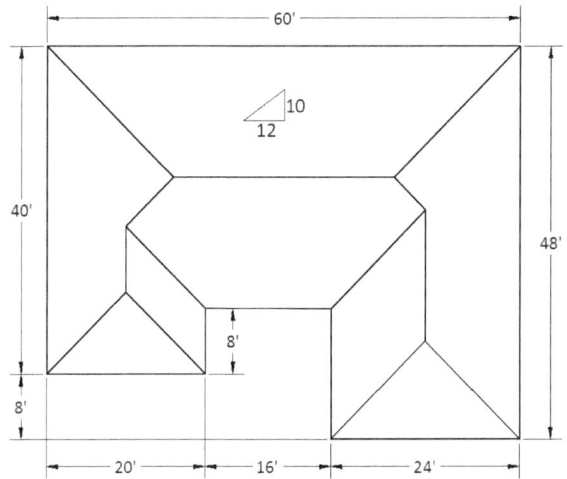

8. HVAC (heating, ventilation, air conditioning) contractors use a WYE (named after its resemblance to the letter Y) to branch off from the main trunk line to get air to each room in a house. To equalize the pressure, the inflow duct of the WYE should ideally have the same area as the other two outflow ducts combined. Calculate the diameter for a duct that is to branch into 6" and 10" diameter ducts, rounded to the nearest whole number.

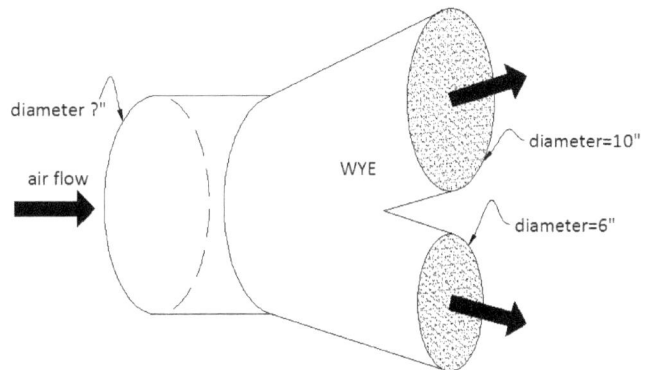

9. HVAC (heating, ventilation, air conditioning) contractors use a boot to transition from the rectangular register (commonly seen on the floor or ceiling of a house) to a circular duct. To equalize the pressure, the rectangular end of the boot should, ideally, have the same area as the circular end. Calculate the diameter for a circular end that will match a 6" x 14" rectangular end, rounded to the nearest whole number.

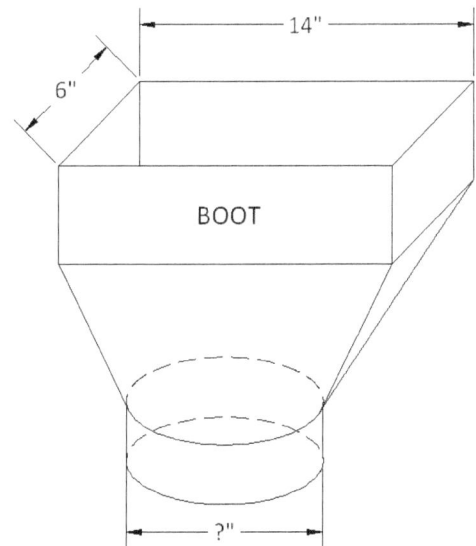

10. Find the total length of belt stretched around three pulleys, each with a 3-inch radius, placed 8", 10" & 11" apart, rounded to one decimal place. Hint: Use the formula for arc length in the appendix.

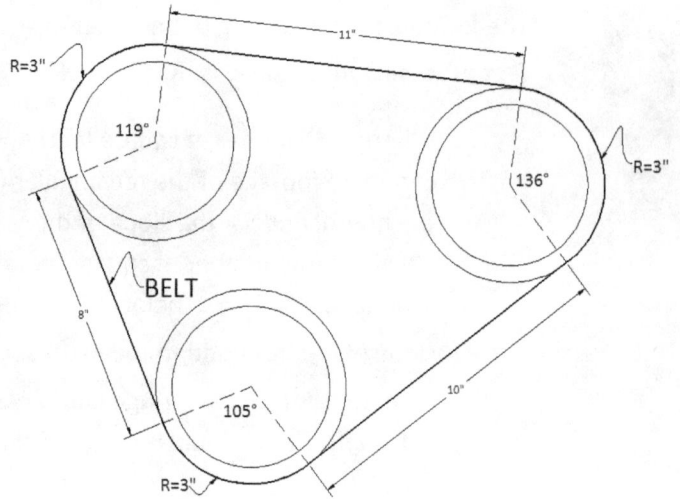

11. Find the area of lawn that a sprinkler waters if it is set to 120 degrees and has a radius of 14 feet, rounded to one decimal place.

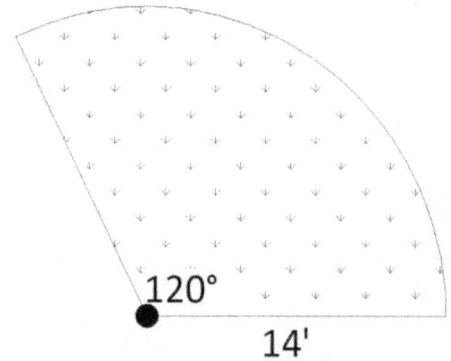

12. Find the number of square yards of carpet that are needed for a rectangular room that is 26' wide by 34' long, rounded up to the nearest square yard.

13. Use the steel design table in the appendix to find the weight of the welded steel beam to the nearest pound.

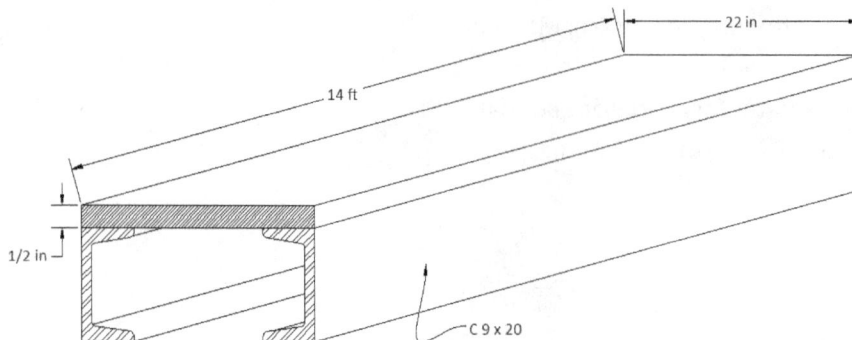

14. Use the steel design table in the appendix to find the weight of the welded steel platform to the nearest kilogram.

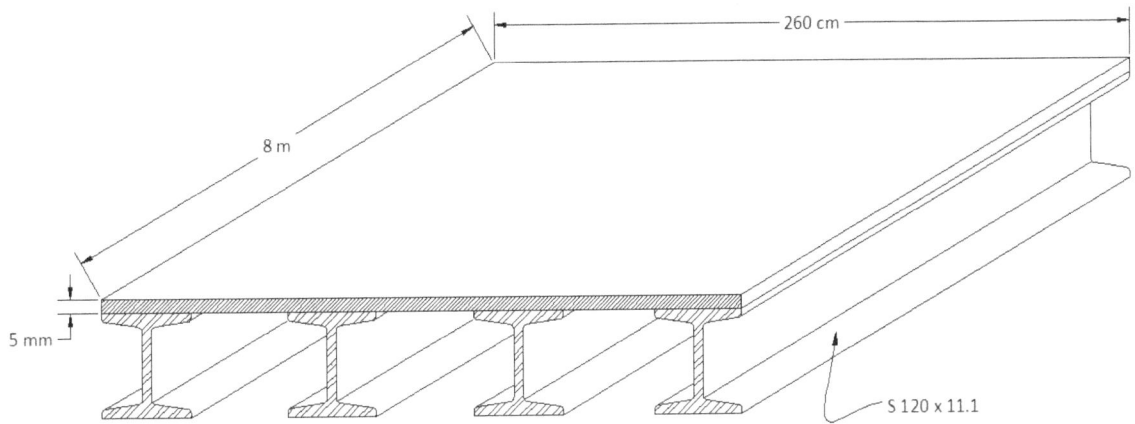

260 cm

8 m

5 mm

S 120 x 11.1

15. Use the steel design table in the appendix to find the weight of the welded steel base and post to the nearest pound.

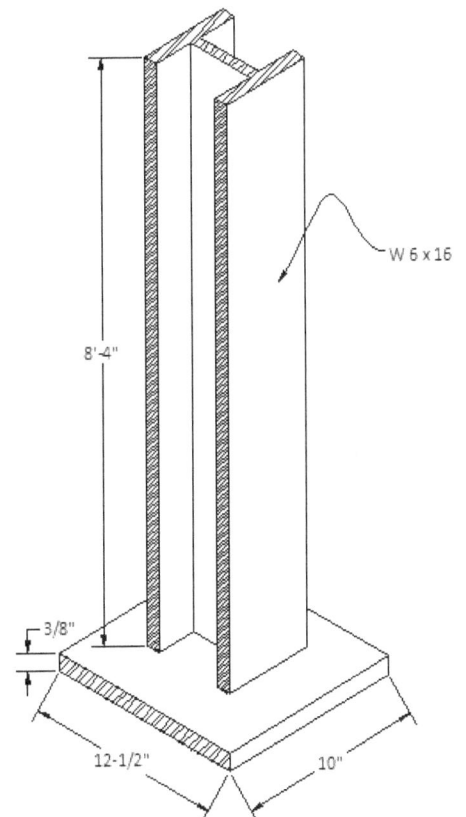

W 6 x 16

8'-4"

3/8"

12-1/2"

10"

97

16. Determine the area of scrap (shaded area) that would result from cutting the parallelogram from the trapezoid as shown in the figure. Answer in square centimeters.

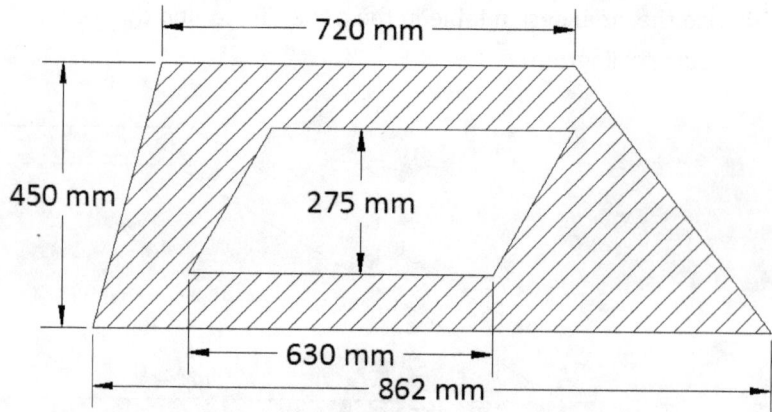

720 mm

450 mm

275 mm

630 mm

862 mm

17. **Challenge Problem:** The sheet metal container below is to be covered in a rust proof coating. How much surface area should be considered if the entire container is to be coated inside and out?

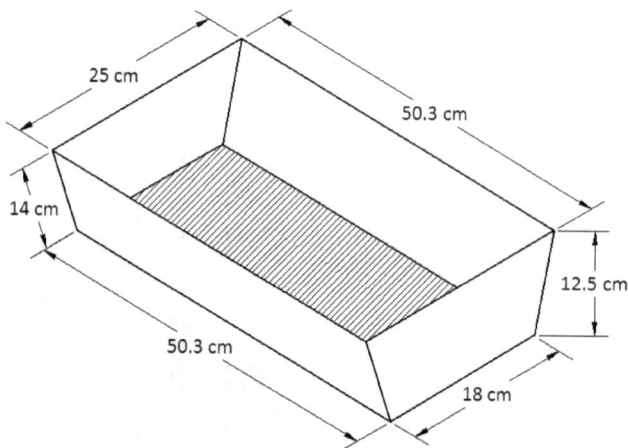

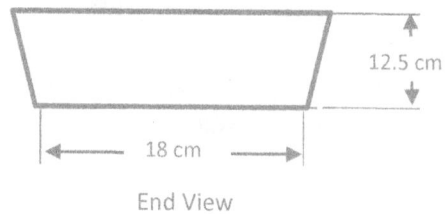

25 cm

50.3 cm

14 cm

12.5 cm

50.3 cm

18 cm

12.5 cm

18 cm

End View

18. The part in the diagram was cut from a steel plate measuring 100 cm x 40 cm. Find the area of the shaded part.

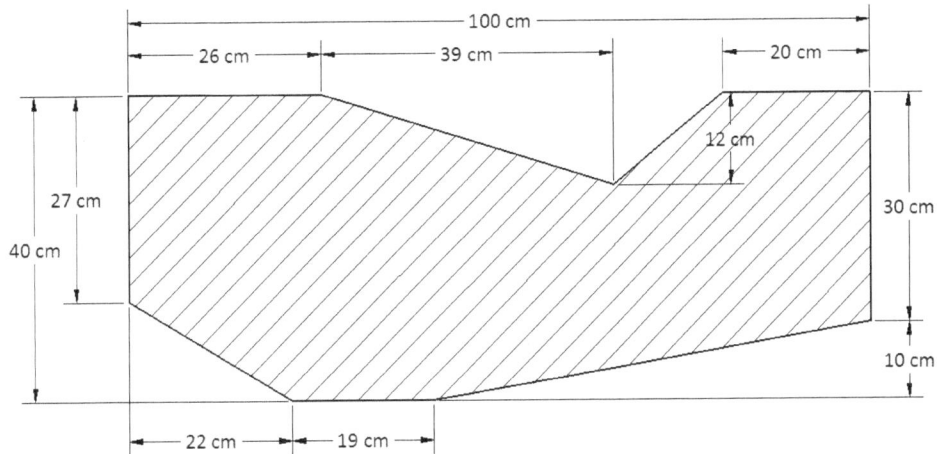

19. The pipe in the drawing is supported below a beam by hanging it with strapping. Find the length of strapping needed, rounded to the nearest $\frac{1}{8}$ inch.

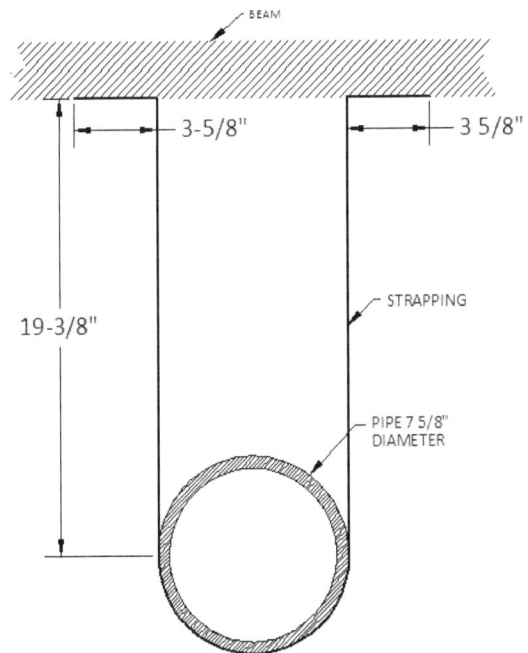

20. The pipe in the drawing is supported in a corner by hanging it with strapping. Find the length of strapping needed, rounded to the nearest inch, to make 18 supports.

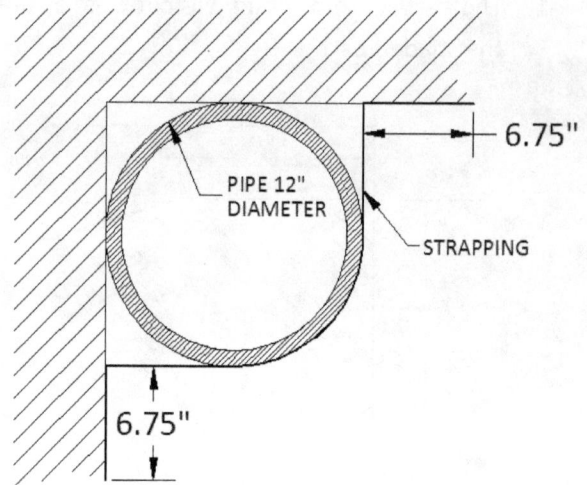

PIPE 12"
DIAMETER

6.75"

STRAPPING

6.75"

21. **Challenge Problem:** Determine the perimeter of the part to the nearest $\frac{1}{1000}$ inch.

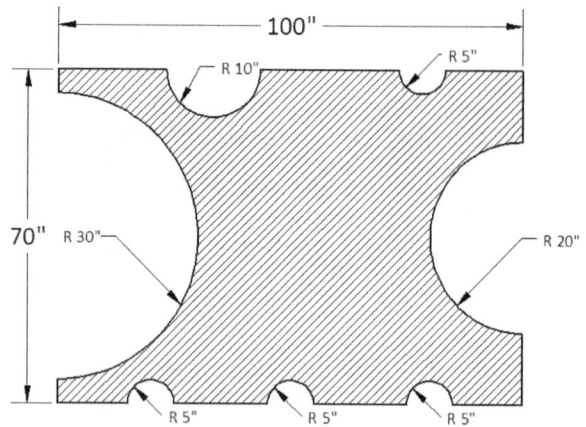

100"
R 10"
R 5"
70"
R 30"
R 20"
R 5"
R 5"
R 5"

22. **Challenge Problem:** Determine the area of the part to the nearest $\frac{1}{1000}$ square inch.

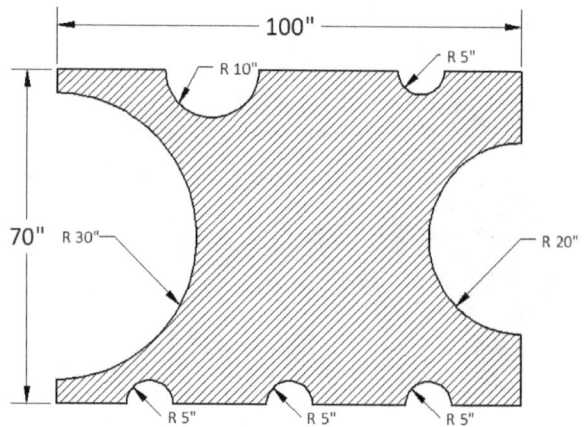

100"
R 10"
R 5"
70"
R 30"
R 20"
R 5"
R 5"
R 5"

23. **Challenge Problem:** The diagram represents one section of a wrought iron fence. Determine the length of material needed for 27 sections like this. Round to the nearest $\frac{1}{100}$ in.

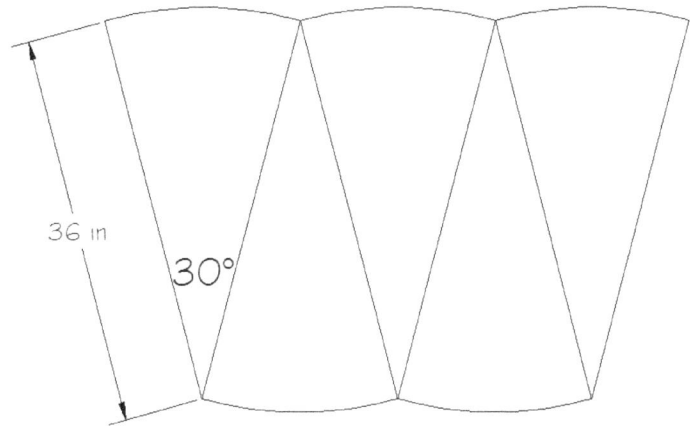

36 in

30°

24. Find the weight of the table frame in the drawing to the nearest $\frac{1}{100}$ pound. (you will need the appendix to find the weight of the feet).

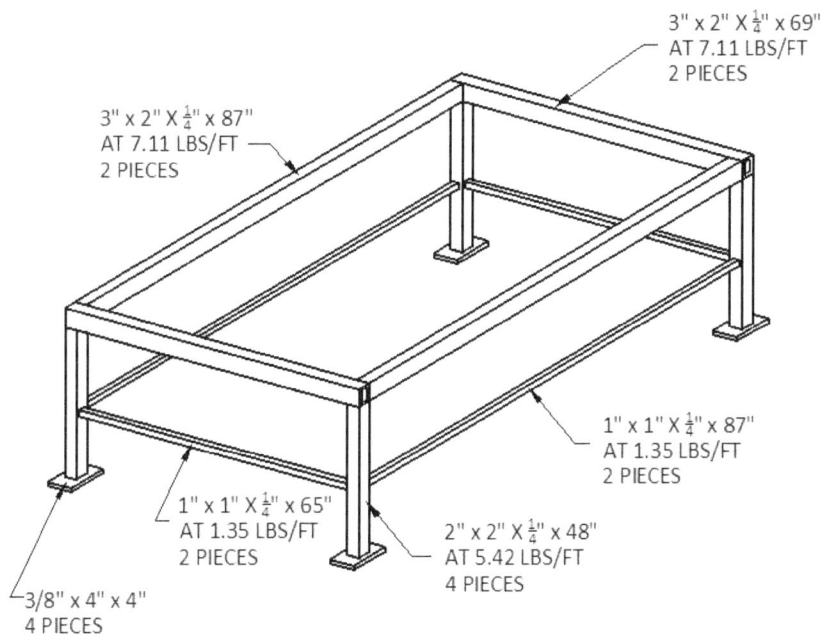

3" x 2" X $\frac{1}{4}$" x 69"
AT 7.11 LBS/FT
2 PIECES

3" x 2" X $\frac{1}{4}$" x 87"
AT 7.11 LBS/FT
2 PIECES

1" x 1" X $\frac{1}{4}$" x 87"
AT 1.35 LBS/FT
2 PIECES

1" x 1" X $\frac{1}{4}$" x 65"
AT 1.35 LBS/FT
2 PIECES

2" x 2" X $\frac{1}{4}$" x 48"
AT 5.42 LBS/FT
4 PIECES

3/8" x 4" x 4"
4 PIECES

3.3: Surface Area

In this section, we will find the surface area of a solid. For example, if we are going to paint a box-shaped room (called a prism), we would have to find the entire surface area of the four walls and the ceiling. Solids such as cylinders, pyramids, cones, and spheres have simple formulas that enable us to determine their surface areas. The skill you gained with the order of operations in section 1.5 will be essential, since some of the surface area formulas are quite complicated. Refer to the appendix for any formulas you may need in this section.

Example 3.3.1: Painting a cylindrical tank

Calculate the number of gallons of paint it will take to cover the top and sides of the steel gas tank. One gallon of paint covers 400 ft^2.

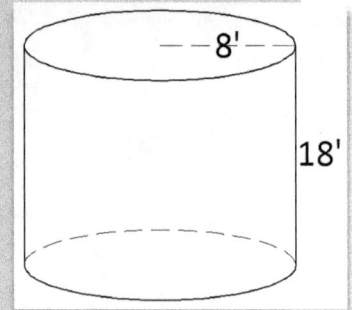

Solution:

The formula in the appendix states that SA = $2\pi r^2 + 2\pi rh$. An understanding of this formula is immediately required. The $2\pi r^2$ is the area for the top and bottom since the area of a circle is A = πr^2.

SA = $\pi r^2 + 2\pi rh$ the formula we will use since we are only covering the top

SA = $\pi 8^2 + 2\pi 8 \cdot 18$ enter the numbers into the formula

SA = $64\pi + 288\pi = 352\pi \approx 1106$ ft^2 simplify

$1106 \text{ ft}^2 \times \dfrac{1 \text{ gal}}{400 \text{ ft}^2} = \dfrac{1106 \text{ gal}}{400} \approx 2.8$ gallon   convert from square feet to gallons

Final Answer: We would need to purchase three gallons of paint to cover the top and sides of the tank.

Use the chart for the weight of plate steel in the appendix to calculate the weight of the table:

Example 3.3.2: Weight of a steel table

Calculate the weight for the $\frac{7"}{16}$ - inch thick rectangular table with semicircular (half circle) ends.

Solution:

The weight in the table states that $\frac{7"}{16}$ thick steel weighs 17.85 lbs/ft^2. We need to calculate the area of the table in square feet.

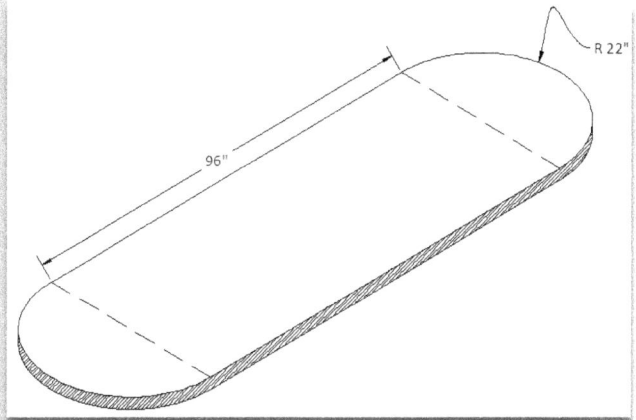

96 in x 44 in = 4224 in^2	the rectangular portion of the area
$A = \pi r^2$	the formula for the area of a circle
$\pi \bullet 22^2 \approx 1520.5$ in^2	substitute the radius into the formula to find the circular portion of the area
5744.5 in$^2 \times \frac{1 \text{ ft}^2}{144 \text{ in}^2} \approx 39.9$ ft^2	total area in square inches converted to square feet
39.9 ft$^2 \times \frac{17.85 \text{ lbs}}{1 \text{ ft}^2} \approx 712$ lbs	total area in square feet converted to pounds

Final Answer: The table weighs approximately 712 pounds. Bring some friends to help lift it!

A hip roof on a house is composed of rectangles and a pyramid. It represents a practical and challenging surface area problem for a carpenter.

Example 3.3.3: Surface area of a roof

Calculate the surface area of the top of the hip roof.

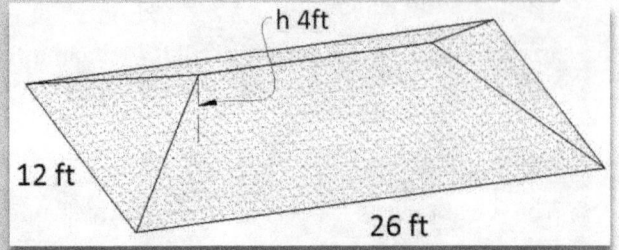

Solution:

A hip roof can be considered to be two half pyramids separated by two identical rectangles, as the illustration implies. First calculate the surface area of the pyramid.

$SA = b\sqrt{b^2 + 4h^2}$ the formula for the surface area of a pyramid (not including the base)

$SA = 12\sqrt{12^2 + 4 \times 4^2}$ enter the numbers into the formula

$SA = 12\sqrt{144 + 64} \approx 173$ ft^2 simplify

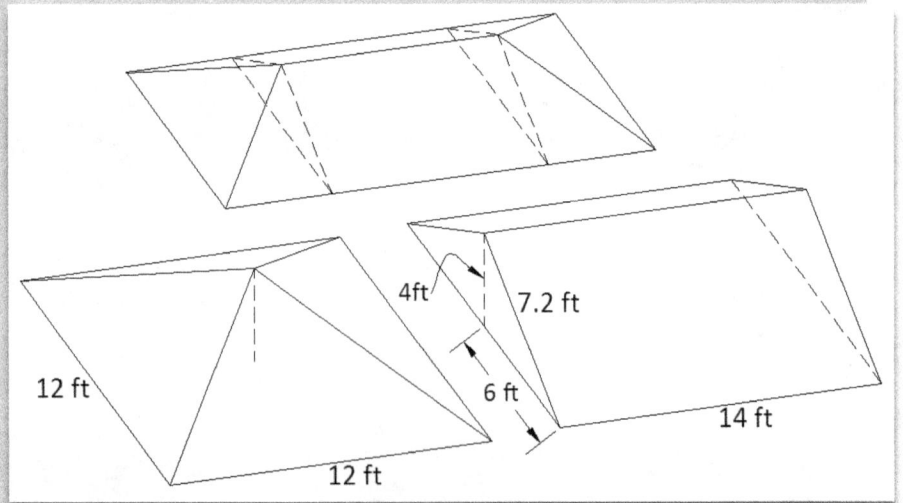

Next calculate the surface area of the rectangular sections.

We need to use the Pythagorean Theorem to find the width of one rectangular portion of the roof.

$\sqrt{6^2 + 4^2} \approx 7.2$ labeled in the diagram above

$14 \times 7.2 \approx 101$ ft^2 the area of one side of the rectangular portion of the roof

Final Answer: Total surface area of the roof $\approx 173 + 2(101) \approx 375$ ft^2.

Note: Another possible technique would be to consider the roof as composed of two identical triangles and two identical trapezoids. Use the area formulas for these shapes (found in the appendix) to arrive at the same answer and decide which method you prefer.

Section 3.3: Surface Area

1. The concrete wall sits on the ground so that the bottom does not need to be coated with sealer. Find the number of gallons of sealer that must be purchased to put two coats on the wall below, rounded up to the nearest whole gallon. Note: One gallon covers 400 square feet.

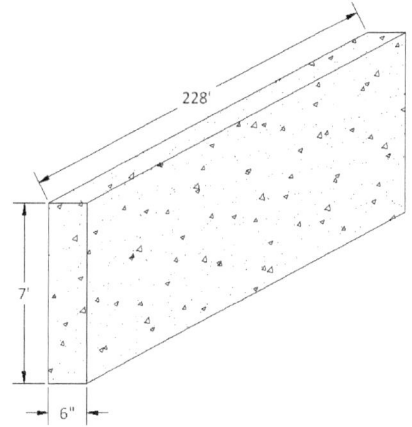

2. Find the surface area of the circular part if its diameter is 198-mm, its thickness is 12-mm, and the diameter of the holes is 42-mm, rounded to the nearest mm^2. Hint: The four holes reduce the surface area on the top and bottom, but increase the surface area inside (as illustrated by the shading within each hole).

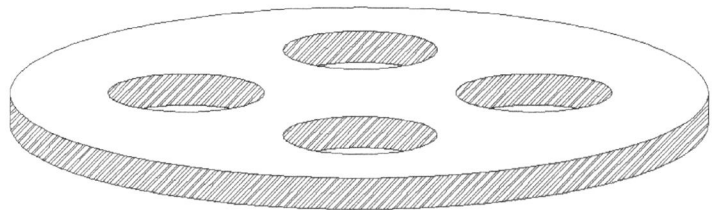

3. If the 24' x 12' x 10' wall that is six inches thick is to be coated with stucco, calculate the number of square feet of stucco required for the inside, outside and top of the wall.

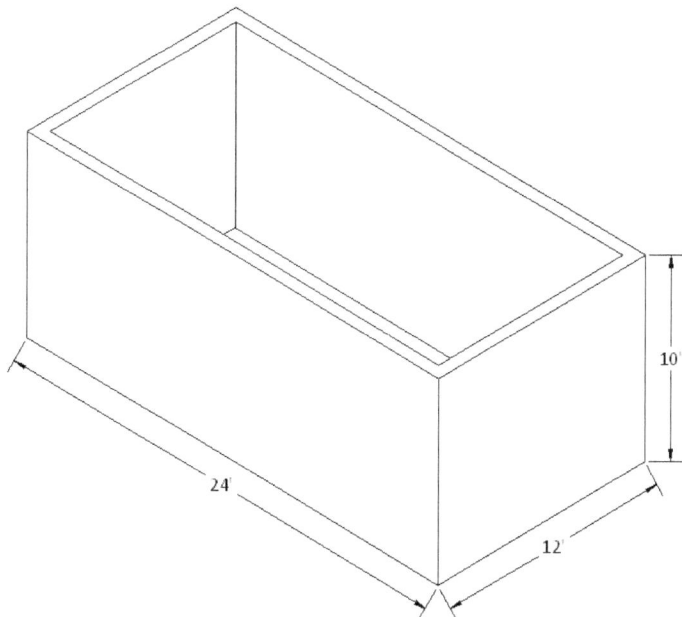

4. Find the total surface area of the washer, rounded to one decimal place. Hint: Think of the washer as a cylinder through which a hole has been drilled.

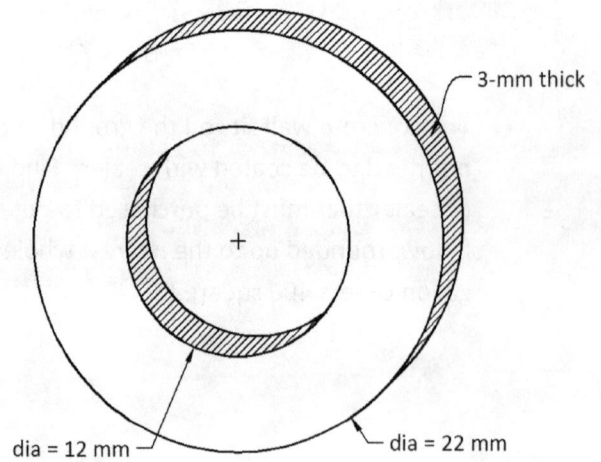

3-mm thick

dia = 12 mm
dia = 22 mm

5. Find the total surface area of the propane tank, rounded to one decimal place. Hint: Think of the tank as a cylinder with a half sphere at each end.

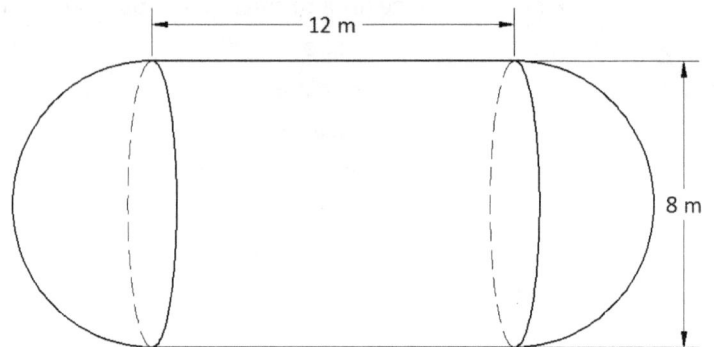

12 m

8 m

6. A welder is building the storage container shown below with sides, bottom and center divider made of 10-mm plate steel. Use the steel design table in the appendix to calculate the weight of the box, rounded to the nearest kilogram.

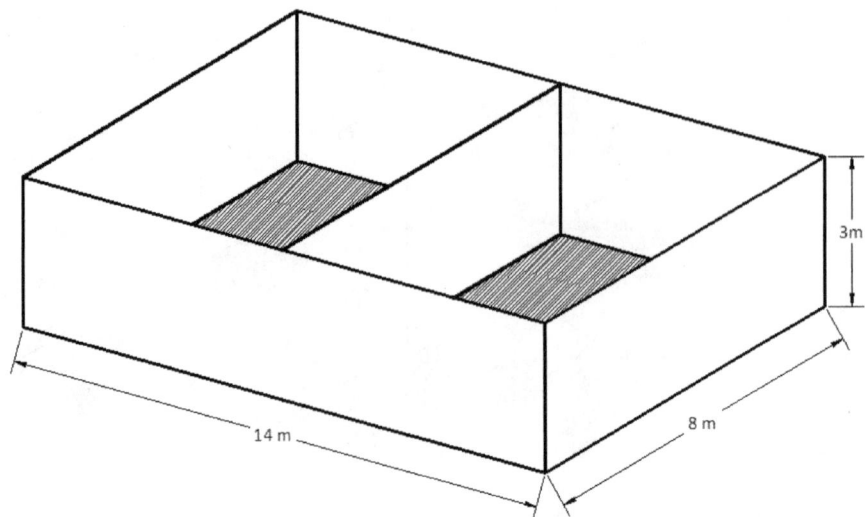

3m

14 m

8 m

7. A welder is building a hollow water storage tank made of $\frac{3"}{8}$ plate steel dimensioned as shown in the diagram. Use the steel design table in the appendix to calculate the weight of the tank, rounded to the nearest pound.

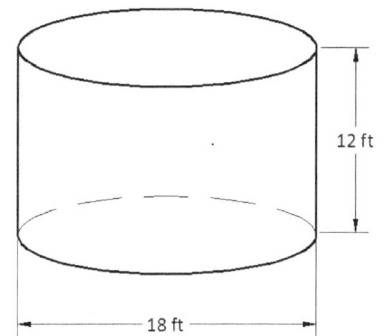

12 ft

18 ft

8. A manufacturer begins with 28-cm x 16-cm rectangular pieces of plate steel 8-mm thick. The corners are rounded off with a 2-cm radius and a 3-cm radius hole is drilled in the center. Use the steel design table in the appendix to calculate the weight of 540 of the finished plates, rounded to the nearest kilogram for shipping purposes. Note: Consider the units of measure since they do not match the units in the steel design table.

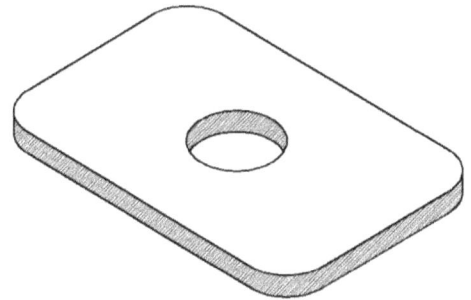

3.4: Volume

In this section we will find the volumes for the solids that we considered in section 4.2. Again the skill you gained in section 1.5 with the order of operations will be essential. Refer to the appendix for the formulas. Skill with volume allows a tradesman to calculate the weight of a hot tub, order the right amount of concrete for a foundation, and determine the capacity of a combustion engine.

Speaking of combustion engines, perhaps you have your eye on a red BMW F 800 GS enduro motorcycle with a 2-cylinder water cooled engine. Car and motorcycle engines are classified by the volume (capacity) of the cylinders in which the motion-generating explosions occur. Notice in the engine specifications below that the F 800 GS has two cylinders with an 82-mm bore (diameter), a 75.6-mm stroke (height), and a capacity of 798 cc (which BMW generously rounded up to 800 for the name). The volume of a cylinder is $V = \pi r^2 h$. For the BMW, one cylinder would be $\pi \cdot 41^2 \cdot 75.6 \approx 399{,}245$ mm^3. Fortunately we are not done, enjoying math as we do, since our answer is in mm^3 and motors are measured in cm^3 (often abbreviated cc).

Relying on our skill with dimensional analysis, $399{,}245$ mm$^3 \times \frac{1 \text{cm}^3}{1000 \text{mm}^3} \approx 399.2$ cm^3.

There are two cylinders of this size so the capacity is 399.2 x 2 $\approx$ 798.4 cm^3.

Engine

Type	Water-cooled, 2-cylinder, 4-stroke, four valves per cylinder, two overhead camshafts, dry sump lubrication
Bore x stroke	82 mm x 75.6 mm
Capacity	798 cc
Rated output	85hp (63 kW) at 7,500 rpm
Max. torque	62 lb/ft (83 Nm) at 5,750 rpm
Compression ratio	12.0 : 1

Side Problem: My 1965 Chevy truck has a 292, 6-cylinder engine. Being an American vehicle, the 292 capacity refers to cubic inches and it has 6 cylinders. Apply the equation-solving skills you gained from chapter 2 to calculate the bore if the stroke is 4.5 inches (rounded to the nearest thousandth of an inch).

Solution: The engine would have a 3.711-inch bore.

Example 3.4.1: Concrete Foundation

Calculate the number of cubic yards of concrete for the residential foundation wall.

Solution:

The formula in the appendix states that V = LWH for a prism. The foundation is made up of two prisms, and notice that all measurements are in inches except the length of 32 feet.

32 feet = 384 inches.

8 in x 16 in x 384 in = 49,152 in^3 volume for the footing

8 in x 20 in x 384 in = 61,440 in^3 volume for the stemwall

$110{,}592 \text{ in}^3 \times \frac{1 \text{ yd}^3}{46{,}656 \text{ in}^3} \approx 2.4 \text{ yd}^3$ volume for the foundation in cubic yards

Final Answer: The foundation will require 2.4 yd^3 of concrete.

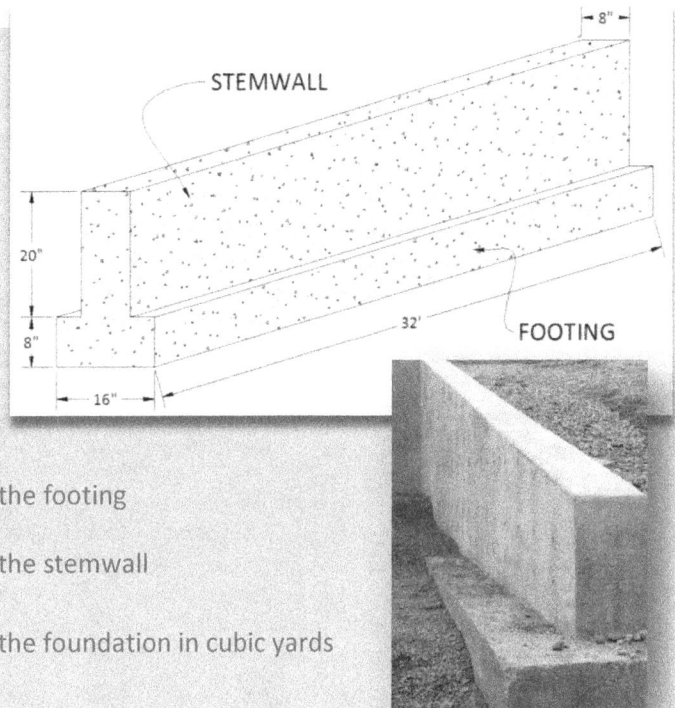

As in the previous example, most real-life volume problems demand a deep enough understanding of the formulas to allow for adjustments. Some solids are made of more than one shape that must be added. Others, like the next example must be subtracted:

Example 3.4.2: Volume of a block

Calculate the volume of the block with four 8-mm diameter holes drilled through it.

Solution:

The formula in the appendix states that V = LWH for a prism.

56 x 42 x 6 = 14,112 mm^3 volume without holes

The volume of a cylinder is V = πr^2h.

$\pi \cdot 4^2 \cdot 6 = 96\pi \approx 301.6$ mm^3 volume of one hole

14,112 − 4(301.6) ≈ 12,905.6 mm^3 volume of 4 cylinders subtracted from the prism

Final Answer: The block has a volume of approximately 12,905.6 mm^3.

Side Problem: Find the surface area of the block in example 4.3.2. Notice this will require subtracting surface area for the missing circles on the top and bottom, but adding surface area for the insides of the holes. Again this will demand a deeper understanding of the meaning of each part of the formula for the surface area of a cylinder.

Solution: The surface area is approximately 6081 mm^2.

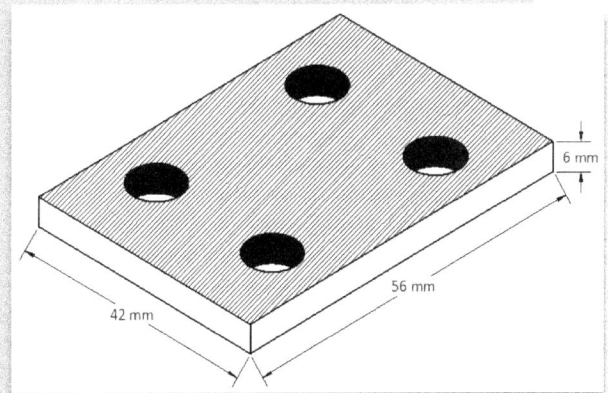

Some volume formulas are quite complex. The frustum (from which we must get our word frustrating) of a cone is such a formula.

Example 3.4.3: Volume of a coffee cup

Calculate the volume of the coffee cup in cm^3 and convert to ounces.

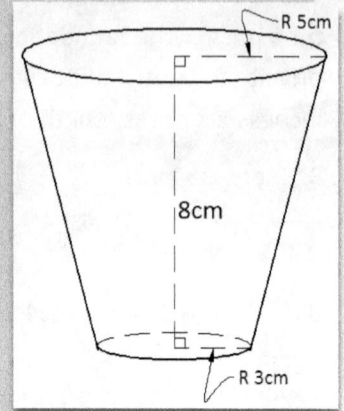

Solution:

$V = \frac{1}{3}\pi h(R^2 + Rr + r^2)$ formula for the volume of a frustum of a cone from the appendix

$\frac{1}{3}\pi \bullet 8(5^2 + 5 \bullet 3 + 3^2) \approx 410.5 \text{ cm}^3$ enter the numbers into the formula and simplify

$410.5 \text{ cm}^3 \times \frac{1 \text{ oz}}{29.574 \text{ cm}^3} \approx 13.9 \text{ oz}$ convert cubic centimeters to ounces

Final Answer: The volume of the cup is approximately 14 ounces.

Side Problem: Find the surface area of the sides and bottom of the cup in example 4.3.3. The formula is modified to $SA = \pi r^2 + \pi(R + r)\sqrt{(R - r)^2 + h^2}$. Since the cup has no top, we do not need πR^2.

Solution: The surface area of the cup is approximately 235.5 cm^2.

Chapter 4
Section 3.4: Volume

1. Find the volume of concrete necessary for the wall in cubic yards, rounded to one decimal place.

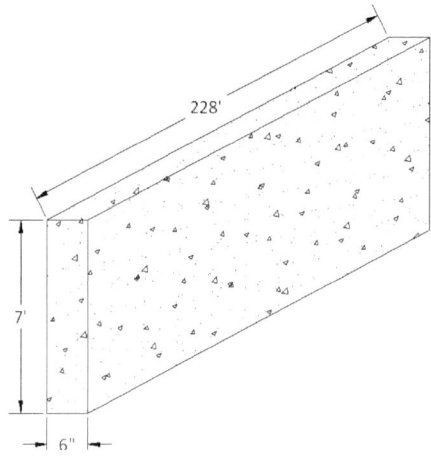

228'

7'

6"

2. Find the volume of concrete necessary to make 34 cylindrical pier pads if r = 14 inches and h = 22 inches. Answer in cubic yards, rounded to one decimal place.

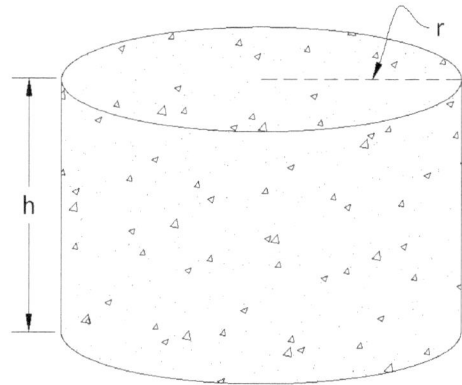

r

h

3. Find the cubic yards of concrete for the sidewalk (top view pictured below), if it is 4 inches thick, rounded to one decimal place. Assume the entire sidewalk is 4 feet wide.

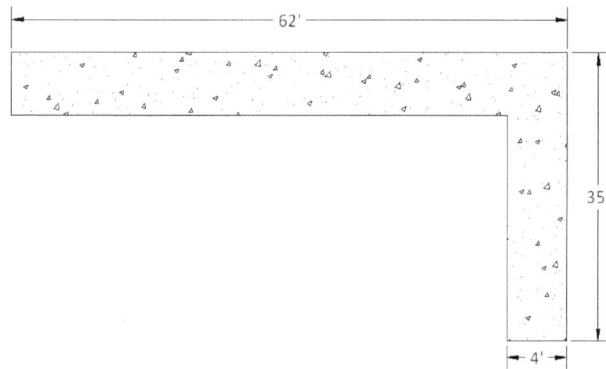

62'

35'

4'

4. Estimate the number of cubic yards of gravel needed to fill a 26' x 20' garage with a layer of gravel that is 18 inches deep, rounded to one decimal place. Hint: Think of the gravel as a box or rectangular prism that is 26' x 20' x 18" (the wall thickness would actually decrease the amount of gravel needed but an estimate can ignore this).

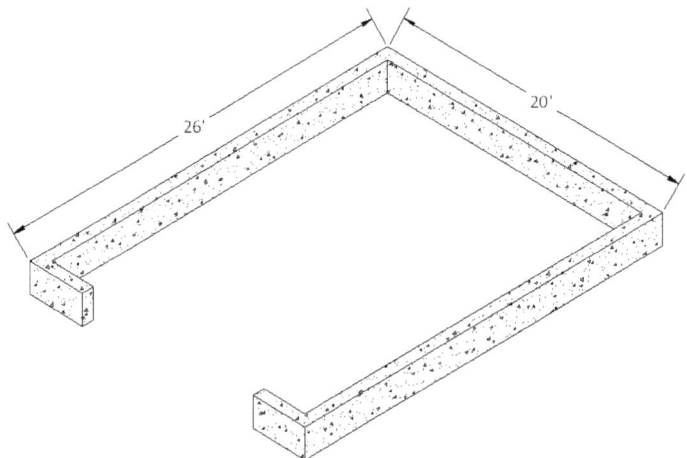

26'

20'

5. Find the number of cubic yards of gravel in the conical pile shown below, rounded to one decimal place.

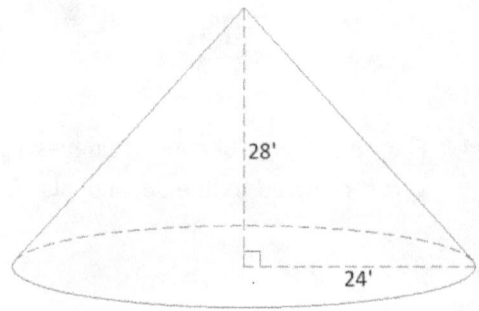

6. Find the volume of the I-beam in cubic inches.

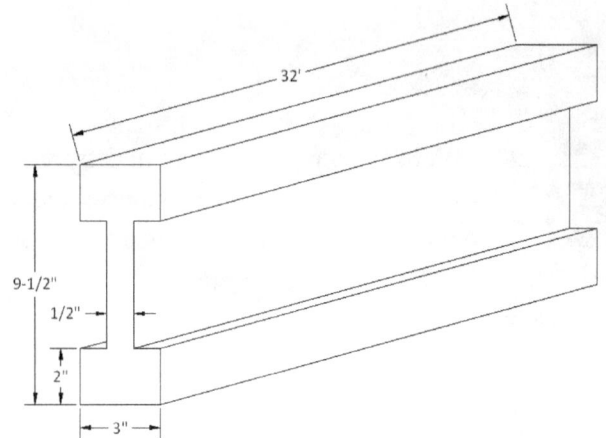

7. Find the volume of the circular part if its diameter is 198 mm, the thickness is 12 mm, and the diameter of the holes is 42 mm. Express your answer in cubic millimeters, rounded to one decimal place.

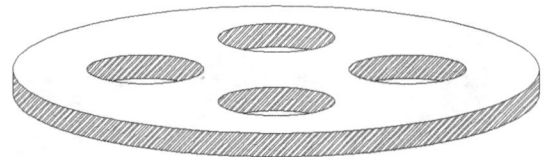

8. Contractors often put shale down as a base layer for a house. To allow room for the foundation crew to work, the base layer is made three feet wider than the footprint of the house in every dimension. Calculate the number of cubic yards of shale necessary for the base layer if it is to be 20 inches deep and three feet wider as illustrated by the dashed line below. Round your answer to one decimal place.

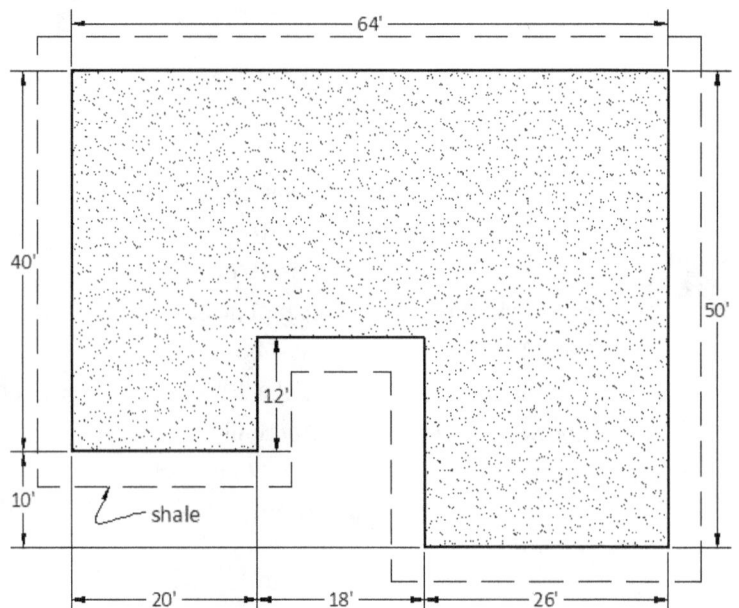

9. One-story foundations are made of a 14" x 6" footing and a 6" x 22" stem wall. Calculate the number of cubic yards of concrete necessary for the foundation below, rounded to the nearest cubic yard. Hint: Find the perimeter of the house and consider the footing and stem wall as separate rectangular prisms of concrete with lengths equal to the perimeter of the house.

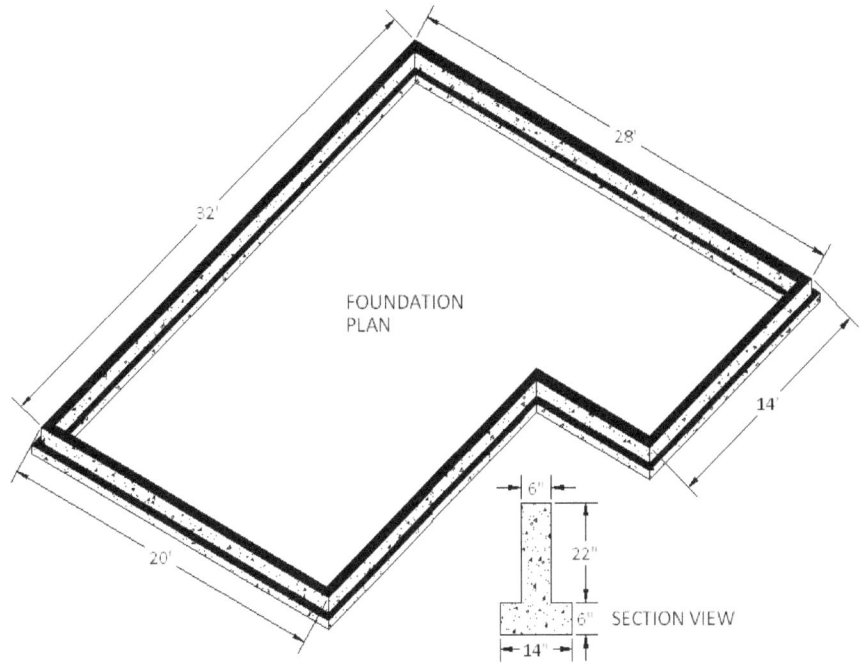

FOUNDATION PLAN

SECTION VIEW

10. Calculate the number of 8" x 8" x 16" concrete blocks required for the 20' x 16' x 8' wall below. Hint: Volume should help. Note: The blocks are actually $\frac{3"}{8}$ smaller to allow for mortar, so the given dimensions are actually the finish dimensions including the mortar joint.

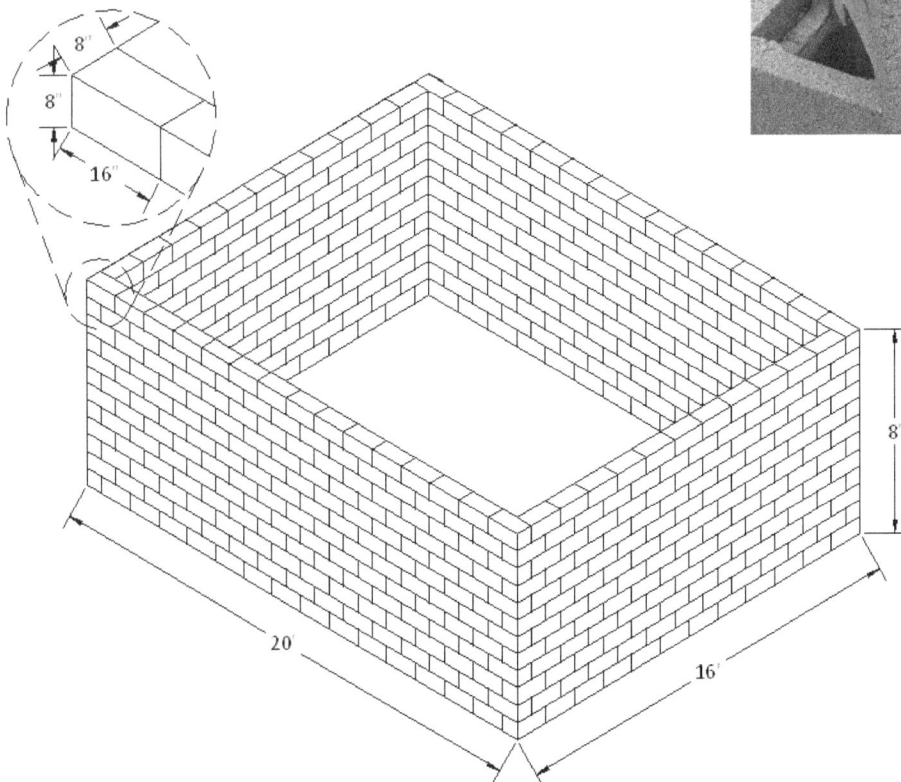

11. Calculate the number of 8" x 8" x 16" concrete blocks required for a 24' x 12' x 10' wall.

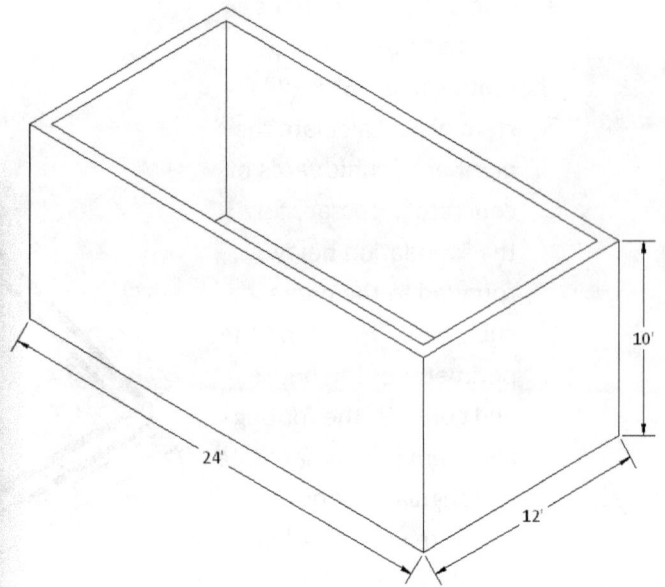

12. Find the total volume of the propane tank, rounded to one decimal place. Hint: Think of the tank as a cylinder with a half sphere at each end.

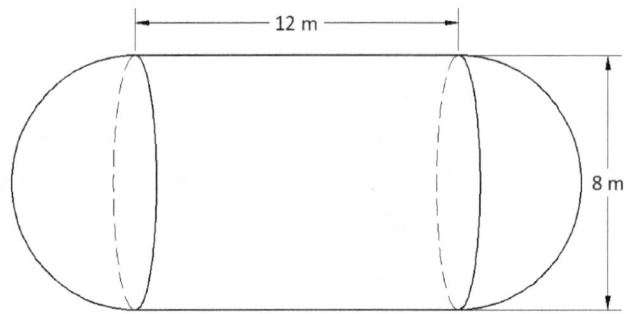

13. Find the total volume of the hip roof in cubic feet, rounded to one decimal place. Hint: Think of the roof as two halves of a pyramid separated by a triangular prism.

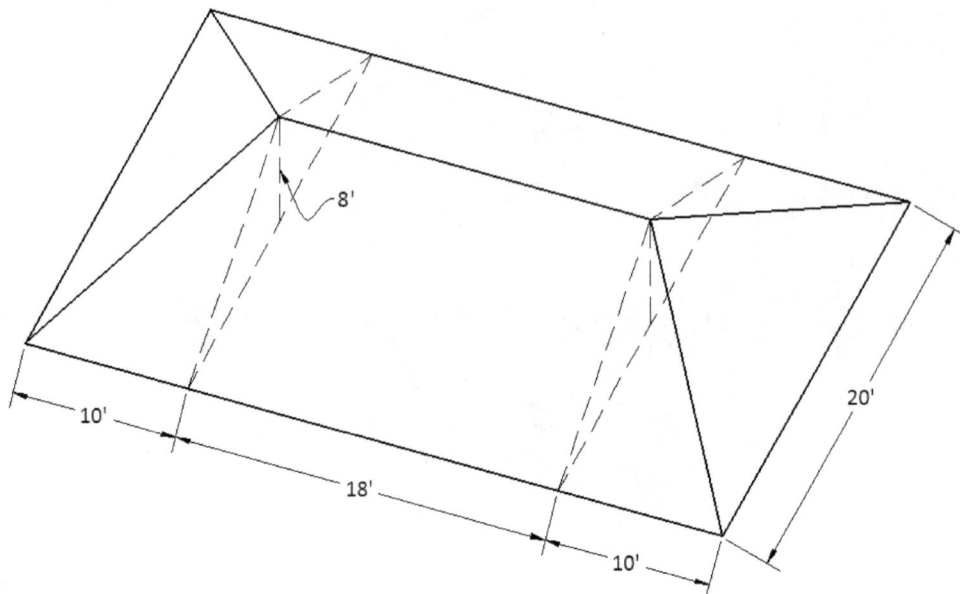

14. Boise Cascade manufactures versa-lam beams (VLB) that weigh 37 lbs. per cubic foot. You have ordered a 5-1/4" x 20" x 32' VLB for a second story floor, calculate its weight so that you can consider how you are going to get it into position. Hint: It is much easier to find the volume of the beam in cubic inches then make a conversion. Round your answer to the nearest pound.

15. The circular steel part is 5 inches in diameter, $\frac{3"}{8}$ thick and has four holes that are each 1 inch in diameter. Find the weight of the part if steel weighs eight grams per cubic centimeter, rounded to the nearest gram.

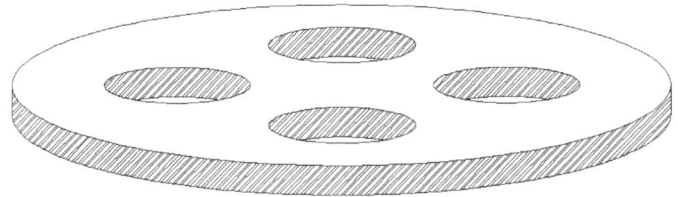

16. Copper pipe is manufactured in three types according to wall thickness. Use the OD (outside diameter) and ID (inside diameter) measurements from the chart to calculate the weight of a pipe rounded to the nearest ounce. The measurements in the chart are in inches and copper weighs 5.16 ounces per cubic inch.

a) Size 1/2   Type L   Length 20'

b) Size 3/4   Type K   Length 10'

c) Size 1 $^1/_2$ Type M   Length 12'

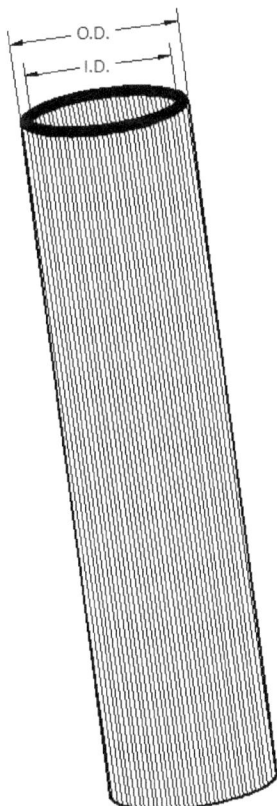

		Type		
		K	L	M
Size	outside diameter (O.D.)	inside diameter (I.D.)		
3/8	1/2	0.402	0.430	0.450
1/2	5/8	0.528	0.545	0.569
5/8	3/4	0.652	0.668	0.690
3/4	7/8	0.745	0.785	0.811
1	1-1/8	0.995	1.025	1.055
1 ¼	1-3/8	1.245	1.265	1.291
1 ½	1-5/8	1.481	1.505	1.527
2	2-1/8	1.959	1.985	2.009

17. A carpenter is building a deck that needs to support a hot tub that weighs 420 pounds empty. Calculate the weight of the hot tub when filled with water, rounded to the nearest 100 pounds. Note: There are 7.48 gallons of water per cubic foot and water weighs 8.345 pounds per gallon of water. Ignore the thickness of the wall.

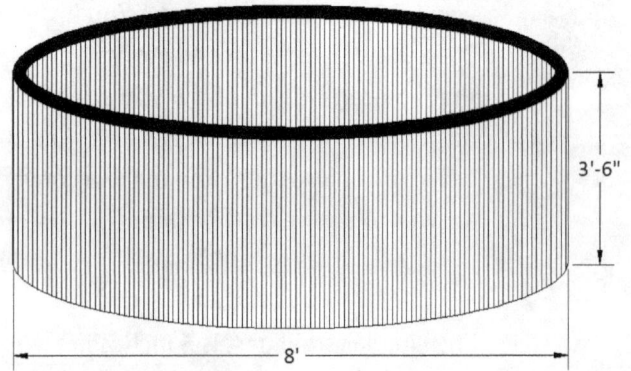

3'-6"

8'

18. A bridge worker wants to transport the concrete wall below and needs to figure out how much it weighs. Calculate its weight, rounded to the nearest 1000 pounds.
Note: Concrete weighs 3915 pounds per cubic yard.

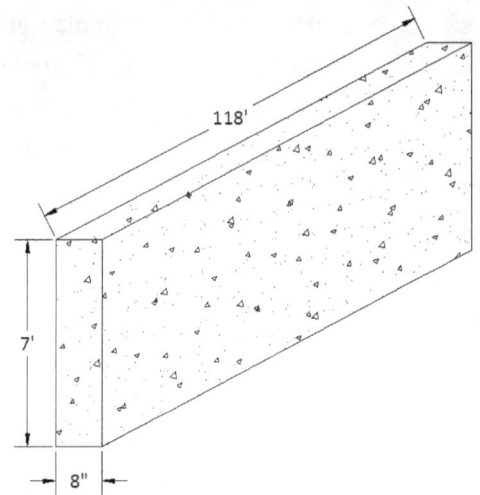

118'

7'

8"

19. Lumber is often sized, for comparison purposes, in board feet. Calculate the number of board feet in the wall if it is made out of 2" x 4" lumber. Round your answer to the nearest board foot.
Note: One board foot is considered to be 12" x 12" x 1".

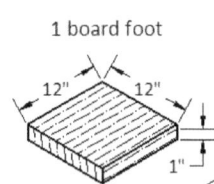

Note: Although the dimensions are actually $1\frac{1}{2}^{"}$ x $3\frac{1}{2}^{"}$, the nominal measurements 2" x 4" are used in the calculation of board feet.

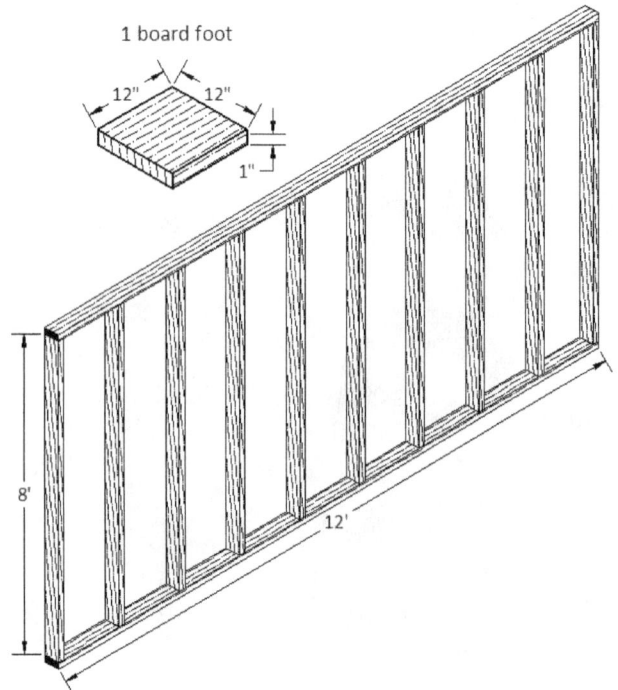

1 board foot

12"   12"

1"

8'

12'

20. Lumber is often priced by the board foot. Calculate the number of board feet in 18 beams that are 4" x 8" x 20'.

    Note: One board foot is considered to be 12"x12"x1".

    Note: Although the dimensions are actually $3\frac{1}{2}$" x $7\frac{1}{4}$", the nominal measurements 4" x 8" are used in the calculation of board feet.

21. A carpenter needs to know the weight of the wall below. Although the lumber is labeled 2" x 4", its dimensions are actually $1\frac{1}{2}$" x $3\frac{1}{2}$" because it is sanded down to maintain accuracy and minimize splintering. Calculate the weight of the 8' wall with 6' studs accurate to the nearest pound.

    Note: Green (wet) Douglas fir weighs approximately 38 pounds per cubic foot.

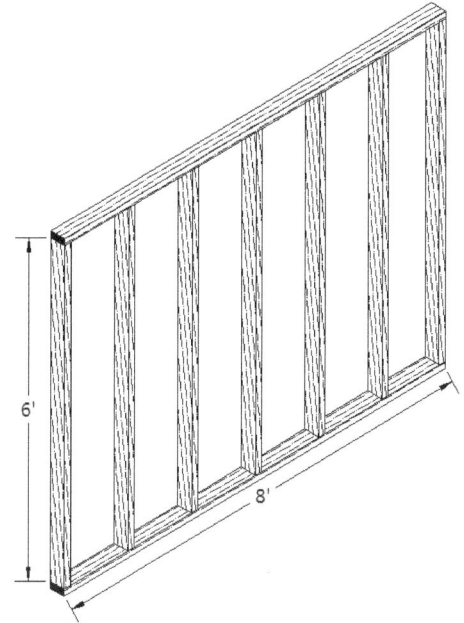

22. Calculate the time it will take to fill the pool below with a hose that flows at 14 gallons per minute (GPM), rounded to the nearest minute.

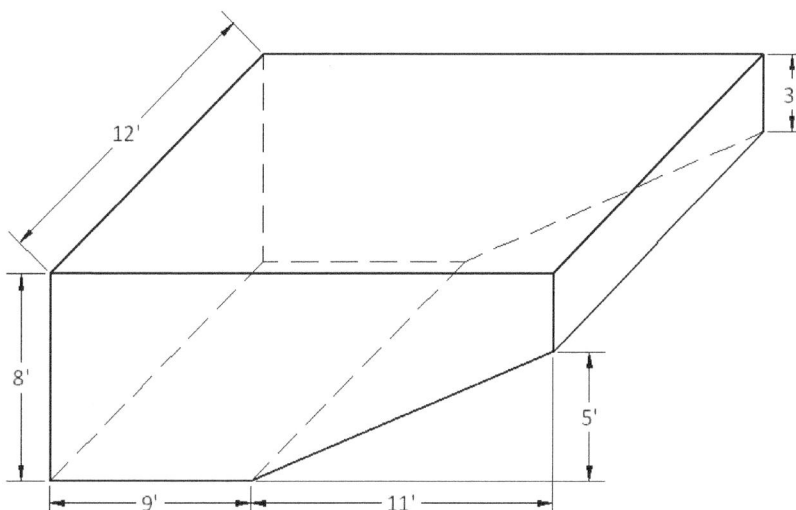

23. **Challenge Problem:** Calculate the diameter size for the five holes to drill in the 1.5-cm thick brass plate to get the weight down to 8850 grams, rounded to one decimal place. Note: Brass weighs 8.4 grams per cubic centimeter.

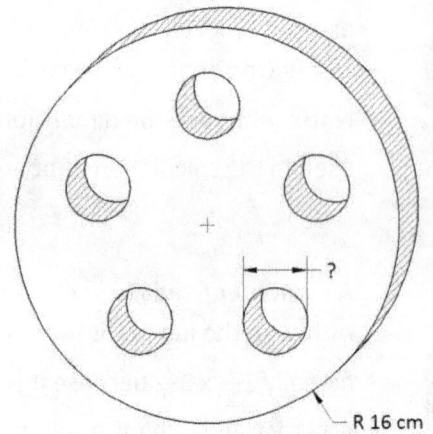

R 16 cm

24. Find the weight of the tank when it is full of water, rounded to the nearest 1000 pounds. The empty tank weighs 3600 lbs, there are 7.48 gallons in one cubic foot, and water weighs 8.345 pounds per gallon. Hint: Think of it as a cylinder with a half sphere at each end.

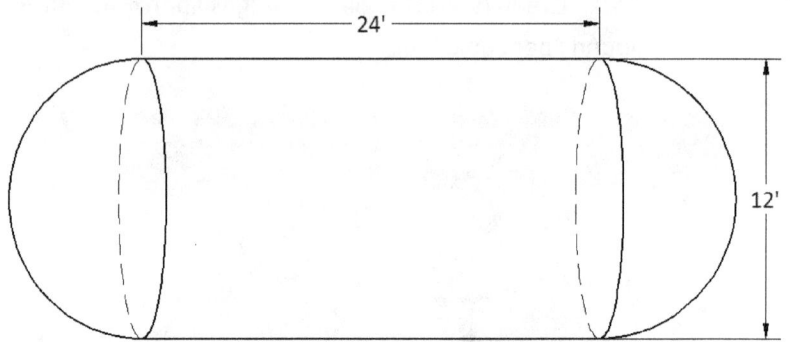

24'

12'

25. Pier pads are made of concrete and used to support decks. Find the weight of 28 pier pads, rounded to the nearest 10 pounds.
Note: The weight of 11.92 cubic inches of concrete is one pound.

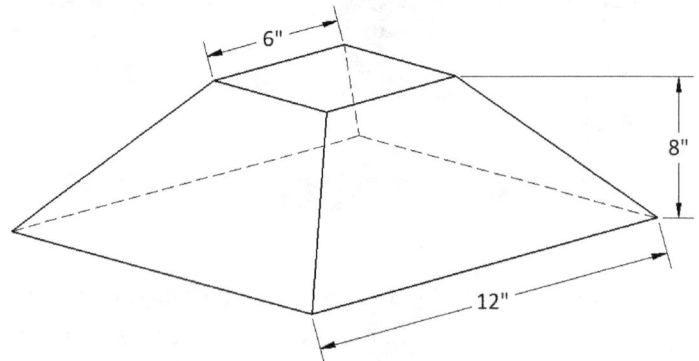

6"

8"

12"

26. Calculate the weight of the titanium plate in ounces, rounded to the nearest ounce. Note: Titanium weighs 2.59 ounces per cubic inch

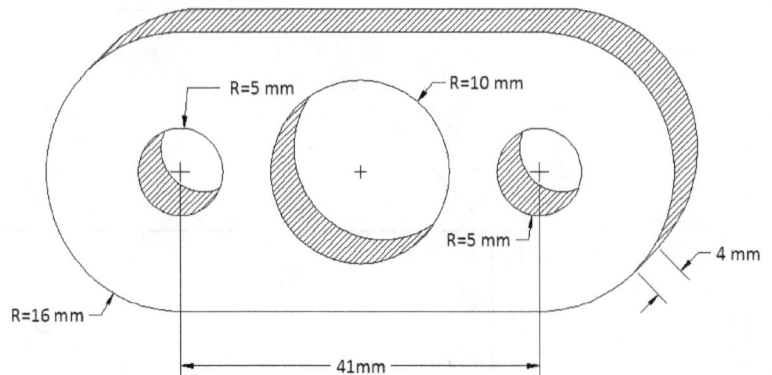

R=5 mm

R=10 mm

R=5 mm

4 mm

R=16 mm

41mm

27. Calculate the weight of the ball (sphere) in grams, rounded to the nearest gram, if it is made of magnesium weighing 1.77 grams per cubic centimeter.

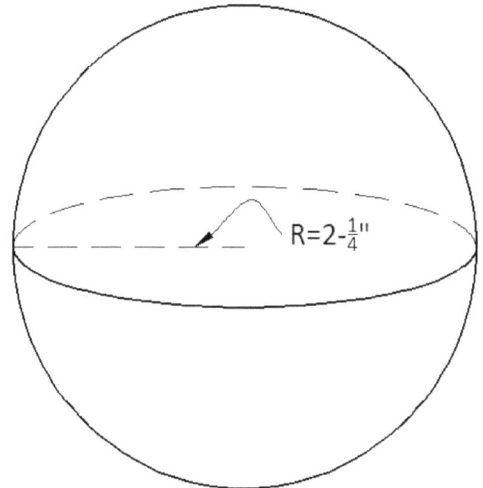

$R=2\text{-}\frac{1}{4}"$

28. Calculate the weight of the object rounded to the nearest pound, if it is made of copper weighing 5.14 ounces per cubic inch. Measurements on the drawing are in inches and the $\varnothing$ symbol in drafting denotes diameter.

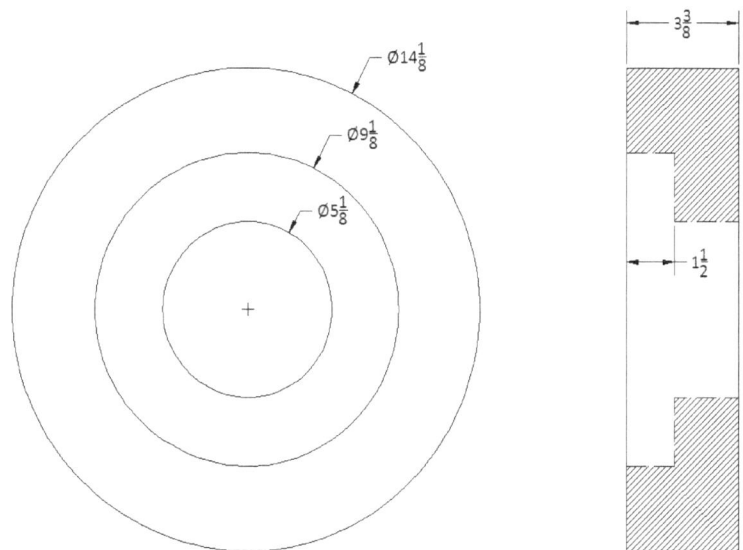

$\varnothing14\frac{1}{8}$

$\varnothing9\frac{1}{8}$

$\varnothing5\frac{1}{8}$

$3\frac{3}{8}$

$1\frac{1}{2}$

29. A welder designs a hollow stainless steel cylindrical tank to the following specifications: 6-foot radius, 14 feet tall, ¼-inch wall thickness. Round your answers to the nearest whole number.

a) Calculate the weight of the empty tank in pounds if stainless steel weighs 4.538 ounces per cubic inch. The tank has a top.

b) Calculate the number of gallons it will hold if one cubic foot will hold 7.48 gallons.

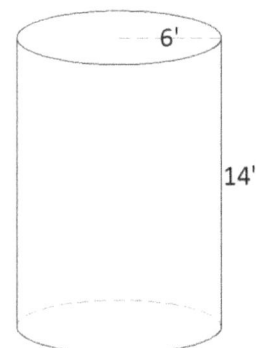

6'

14'

30. The block of steel in the drawing is to be turned on a lathe to a diameter of 220 mm. Calculate the amount of waste to the nearest hundredth of a cm^3.

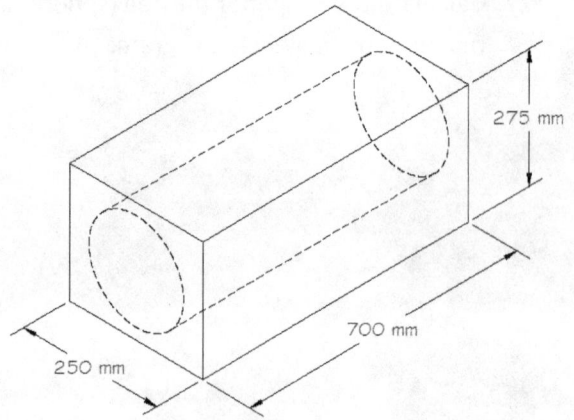

275 mm

700 mm

250 mm

31. Find the volume fo the part made from 2" thick steel. The four small holes have a 2" diameter. The large hole has a 3" diameter. Round to the nearest of in^3. Assume the half cylinder is on top of the plate.

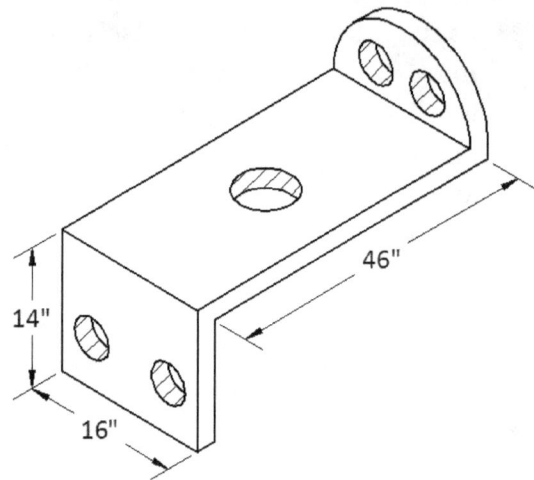

46"

14"

16"

32. Two steel plates are welded together with fillet welds above and below. Find the volume of weld necessary to make 25 double-sided joints like this. Round your answer to the nearest in^3.

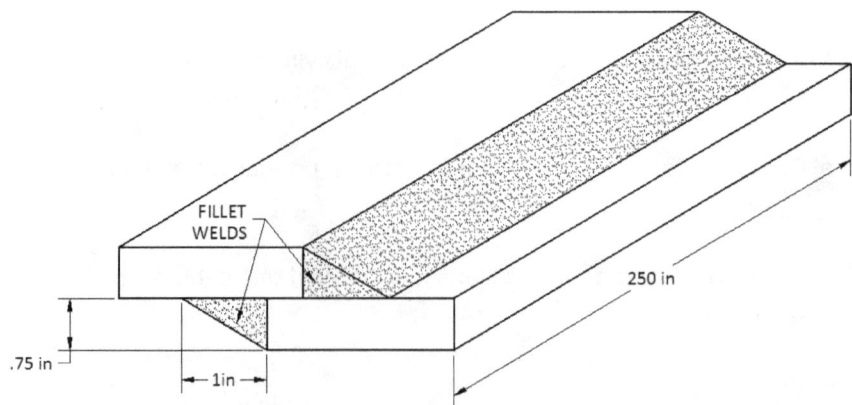

FILLET WELDS

250 in

.75 in

1in

33. A ventilation system in a shop building has an exchange capacity rate of 5500 ft³ of air per minute. Find the time it will it take to exchange all of the air in a shop of the given dimensions to the nearest $\frac{1}{10}$ of a minute.

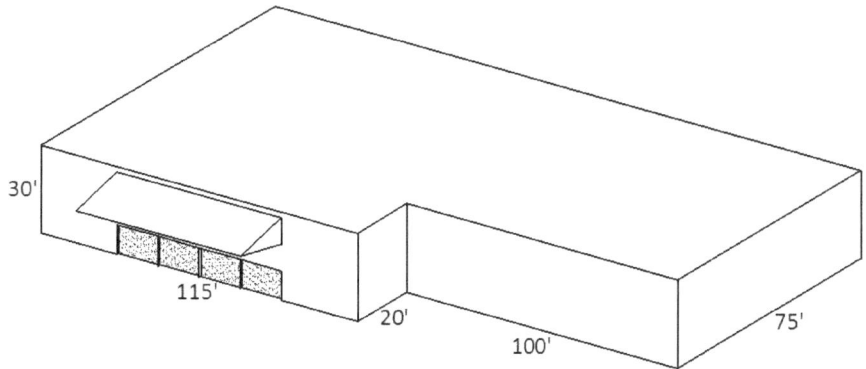

34. Find the diameter of the roll of 1.475 cm thick steel in meters. The material is 6 meters wide and 375 meters long when unrolled. Answer to the nearest $\frac{1}{100}$ of a meter.

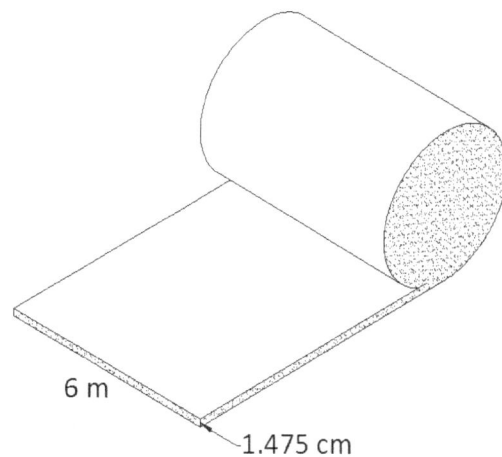

35. Wear on the outer surface of a pipe can be repaired by depositing weld material. Find the volume of weld necessary if .25" is deposited on an 8 foot pipe with a 5.75 inch O.D. (outside diameter). Answer to the nearest $\frac{1}{10}$ of a cubic inch.

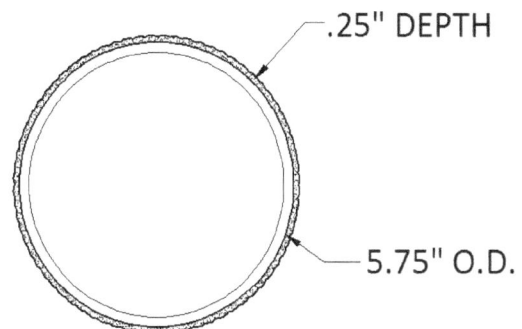

.25" DEPTH

5.75" O.D.

36. Calculate the capacity of the 18" I.D. (inside diameter) pipe to the nearest gallon.

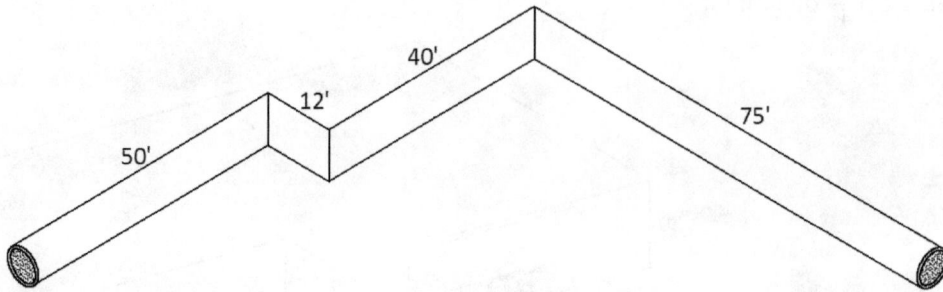

37. **Challenge Problem:** Find the volume of the weld that must be deposited to join two 2800 mm plates based on the dimensions on the drawing. Answer to the nearest cm^3. Drawing is not to scale.

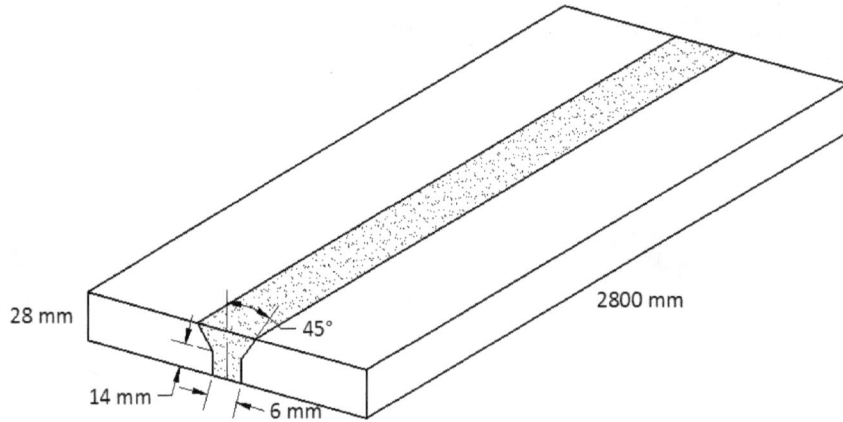

38. **Challenge Problem:** Find the volume of two tanks, including the the pipes, to the nearest gallon. Assume the side and floor thickness of the tanks is 2 inches.

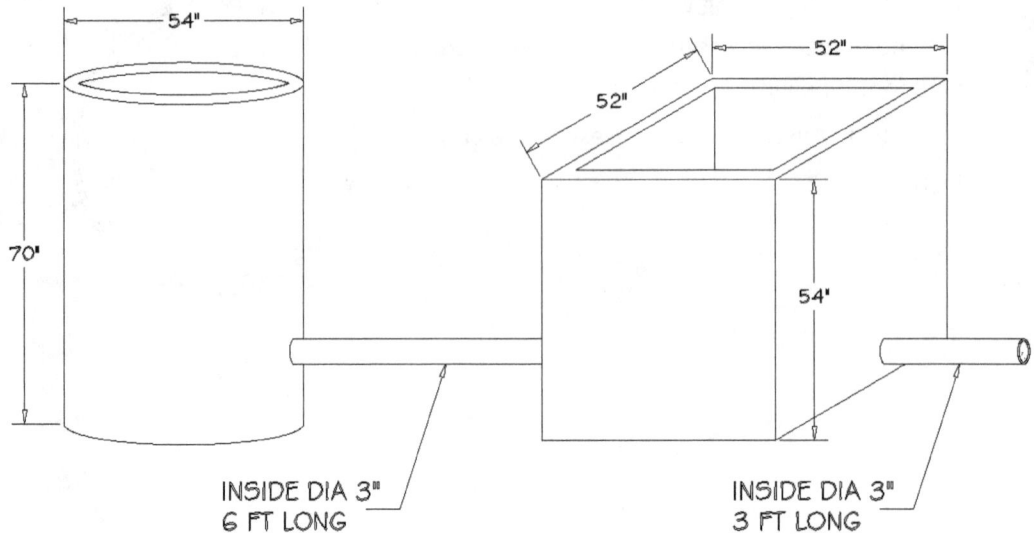

Chapter 4:
Linear Relationships

An excellent place to find, and put to practical use, a myriad of formulas, lies in the study of geometry. Put simply, geometry is the study of two- and three-dimensional shapes. Trailers and railings that a welder builds, cabinet doors and circuit boards that a CNC operator manufactures, and rafters and sheathing that a carpenter cuts, all require geometry. Finding the lengths, angles, area and volume make the production and pricing of these enterprises possible and profitable. In this section, we will draw upon the skills that you gained in the first three chapters to find the surface area and volume of common geometric solids. The goal will not be to simply find the area of a rectangle, but to price and order the paint for a room, not to find the volume of a cone, but to calculate the number of cubic yards of gravel that forms in a conical pile under a conveyor belt. There are a number of formulas, many of which are from calculus, that describe the relationship between a shape's dimensions and its perimeter, surface area, and volume. Refer often to the appendix for the formulas, and add your own notes for clarification when necessary. It is time to apply all that you have learned.

Section 4.1: The Shape of a Linear Equation

Consider the dosage chart below. If we compare the <u>mean weight</u> to the <u>actual dose given</u> it would appear that the dosage increases with weight, as one might expect. A health care provider might use a chart like this to approximate dosage.

Weight Category	< 38 kg	38-47 kg	48-63 kg	64-77 kg	78-97 kg	98-117 kg	118 + kg
Mean weight	29.6	44.2	56.8	70.6	86.0	105.2	122.9
Mean BSA (DuBois method)	1.045	1.370	1.603	1.818	2.015	2.215	2.371
Mean BSA (Mosteller method)	1.045	1.372	1.613	1.841	2.064	2.313	2.535
Actual Dose given	60ml	70ml	80ml	100ml	120ml	140ml	160ml
If dose were 1.416 ml per kg	41.9ml	62.6ml	80.5ml	100ml	121.8ml	149.0ml	174.1ml
If dose were 55.02ml per m² (BSA using DuBois method)	57.5ml	75.4ml	88.2ml	100ml	110.9ml	121.9ml	130.5ml
If dose were 54.30 ml per m² (BSA using Mosteller method)	56.7ml	74.5ml	87.6ml	100ml	112.1ml	125.6ml	137.7ml

http://www.halls.md/ct/dosedata.htm

Two problems present themselves that skill with algebra will solve:

1. What if the patient weight is in pounds instead of kilograms?
2. What if the patient weight is not represented on the chart?

Recall that conversions to other units are easily made applying knowledge of fractions. 29.6 kg would need to be multiplied by a number with units $\frac{lbs}{kg}$ so that kg cancel, leaving our number in pounds. A quick google-search informs us that 1 kg ≈ 2.205 lbs. So, $\frac{29.6 \; kg}{1} \times \frac{2.205 \; lbs}{1 \; kg} \approx 65 \; lbs$. Converting all 7 weights to the nearest pound we find the ordered pairs listed below.

mean weight (lbs)	dose given (ml)
65	60
97	70
125	80
156	100
190	120
232	140
271	160

Note that they are quite logically called "**ordered pairs**". **Pairs**, in that it is a pair of numbers that are related. **Ordered**, in that order matters (giving a 120 pound patient 190 ml of medicine would be a regrettable mistake).

Ordered pairs are commonly written in the form (x,y). By convention the first number is the **independent** variable and the second number is the **dependent** variable. In this example the dose "depends" on the weight of the patient. When representing ordered pairs on a graph the independent variable (x) is assigned to the horizontal axis and the dependent variable (y) to the vertical axis. The "shape" of the data is approximately a line so we call this a **linear relationship**.

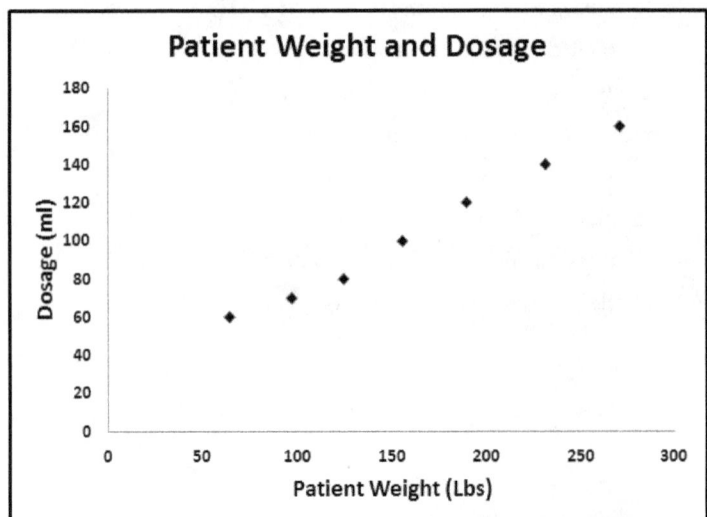

Patient Weight and Dosage

Drawing a "**trend line**" through the points would allow us to solve our second problem and assign a fairly accurate dosage for a weight that is not represented on the chart.

Next consider that the line has **slope**, which turns out to be both interesting and useful.

Slope is defined as the **rise** divided by the **run** so that steeper lines have larger slopes. To find a number for the slope, first choose any two points that represent the line; the rise is the vertical distance between the points and the run is the horizontal distance between the points. The second and the seventh points were chosen in figure 1.1.1. **Important:** *If the trend line does not pass through any of the points, you are free to choose your own.*

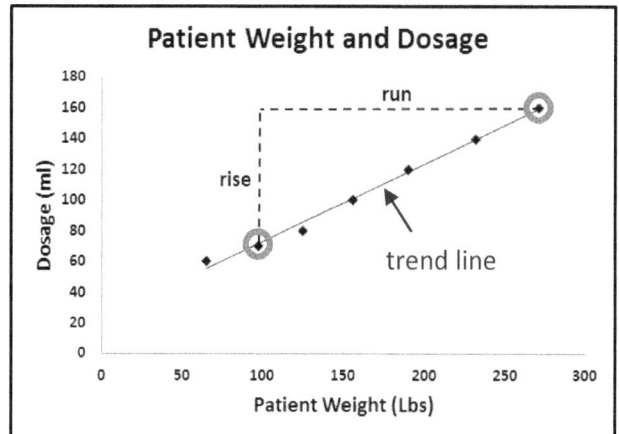

Figure 1.1.1

mean weight (lbs)	dose given (ml)
65	60
97	70
125	80
156	100
190	120
232	140
271	160

Rise, in our dosage example, is found by subtracting the dosages (160ml - 70ml = 90ml).   Run is found by subtracting the weights (271 lbs - 97 lbs = 174 lbs).

So the slope is $\frac{90}{174} = \frac{15}{29} \approx .52$.

With the units included it has a useful meaning. Since $.52 = \frac{.52 \text{ ml}}{1 \text{ lb}}$, we can **add** about .5 ml to our dosage for every **increase** of 1 pound. It is also true that we can **subtract** about .5 ml for every **decrease** of 1 pound since $\frac{.52}{1} = \frac{-.52}{-1}$.

Formally then, slope $= \frac{\text{rise}}{\text{run}} = \frac{y_1 - y_2}{x_1 - x_2}$ **; but practically it refers to the** <u>rate of change</u> **of one quantity relative to another.**

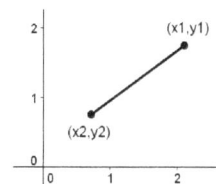

It is interesting to note that choosing different ordered pairs will give you a slightly different slope since the seven points are not in a perfect line. Choosing the 1st and the 5th, slope $= \frac{120-60}{190-65} = \frac{60}{125} = \frac{12}{25} = .48$.

Consider a practical slope (rate of change) example involving car value:

Example 4.1.1: Calculating slope

Find and interpret the slope for the car value from the table between the years 1998 & 2005.

year	car value
1998	$27,500
2001	$22,200
2005	$15,700
2007	$12,800
2013	$2,800

Solution:

$$\text{slope} = \frac{15700-27500}{2005-1998} = \frac{-11800}{7} \approx \frac{-1686}{1}$$

Meaning: a **decrease** of approximately $1686 in value for every **increase** of one year in age.

Note: It does not matter which order you subtract as long as your subtraction starts with x and y from the same point. The other option: slope $= \frac{27500-15700}{1998-2005} = \frac{11800}{-7} \approx \frac{1686}{-1}$.

Meaning: an **increase** of $1686 in value for every **decrease** of one year in age. Numerically, the slope is -1686 in either case.

As demonstrated in figure 1.1.1, a useful graph adheres to the following conventions:

1. independent variable (x) on the horizontal axis
2. dependent variable (y) on the vertical axis
3. a title
4. labels (including units) and a scale for each axis

Consider a practical graphing example involving the same car values:

Example 4.1.2: Graphing

Create a usable graph for the car value from the table. Include a trend line and give a reasonable value for the year 2009.

year	car value
1998	$27,500
2001	$22,200
2005	$15,700
2007	$12,800
2013	$2,800

Solution:

(a) Horizontal Axis - Year is the independent variable since the value depends on the year. There are 15 years between 1998 and 2013. Our graph paper does not have 15 lines so the scale will be 2 years.

(b) Vertical Axis - There is $24,700 between $2,800 and $27,500. $24,700 divided by 14 lines is $1764 which can be rounded up to $2000 for a scale that will be easy to read.

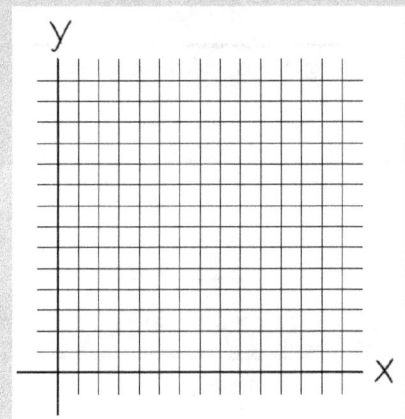

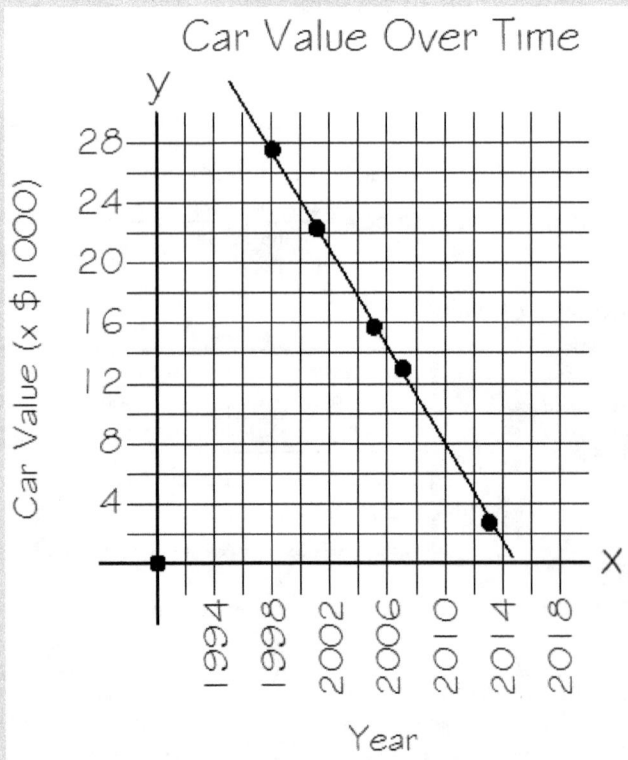

Car Value Over Time

We do not know until the graph is done that the shape of the graph is a line. It would not be at all unusual to have found it curved.

Final Answer: Reading the graph, the car looks to be worth a little under $10,000 in 2009.

Note: Choosing the scale for a graph is a dynamic process and there is often more than one right way of doing it.

Note: The x & y **intercepts** are often interesting, in this case the car's value looks to reach zero in 2015 (the point (2015,0) is the x-intercept).

Section 4.1: Problem Set

1. The Kilowatt-hours (KWH) of electricity a home uses each month are dramatically affected by the temperature difference inside versus outside measured in heating degree days (HDD),(*if it is 60 degrees outside and you heat your house to 72 degrees for 5 days that is 12x5=60 degree days*).

   a) Make a graph of the data (use graph paper, label completely, and choose the correct axis for the independent(x) and dependent(y) variables).

   b) Find the slope between the 133 and 561 degree days, accurate to 2 decimal places.

   c) Add a trend line to the graph.

   d) Choose two representative points from your trend line and find the slope. Explain the meaning of the slope in context.

Heating Degree Days (HDD)	KWH Used
714	1799
469	1269
386	1160
266	860
133	804
62	696
7	677
5	666
87	736
354	1071
561	1602
968	2060

2. A forester takes measurements in a grove of trees. Naturally larger diameter trees also have larger volumes.

   a) Make a graph of the data large enough to include a 13 cm diameter (use graph paper, label completely, let diameter be the independent(x) variable and volume be the dependent(y) variable).

   b) Find the slope between the smallest and largest diameter trees, accurate to 1 decimal place.

   c) Add a trend line to the graph.

   d) Choose two representative points from your trend line and find the slope. Explain the meaning of the slope in context.

   e) Use your trend line to estimate the volume of a 13 cm diameter tree.

Diameter (cm)	Volume (m^3)
7.2	143
7.4	157
7.5	168
7.7	178
7.8	189
8.0	199
8.2	212
8.5	221
8.7	237
8.8	248
9.0	261
9.3	272
9.4	278
9.6	288
9.8	299
10.1	304
10.3	318
10.4	327
10.6	340
10.9	349

3. A forester takes measurements in a grove of trees. Naturally older trees also have larger volumes.

    a) Make a graph of the data larger enough to include a 50 year old tree (use graph paper, label completely, let age be the independent(x) variable and volume be the dependent(y) variable).
    b) Find the slope between the youngest and oldest trees, accurate to 2 decimal places.
    c) Add a trend line to the graph.
    d) Choose two representative points from your trend line and find the slope. Explain the meaning of the slope in context.
    e) Use your trend line to estimate the volume of a 50 year old tree.

Age	Volume (m³)
21	143
22	157
23	168
24	178
25	189
26	199
27	212
28	221
29	237
30	248
31	261
32	272
33	278
34	288
35	299
36	304
37	318
38	327
39	340
40	349

4. A study conducted in the early 1960's in 9 counties in northern Oregon and southern Washington related exposure to the Hanford nuclear plant and mortality rate. Exposure is an index number assigned to a county based on its population and proximity to the contaminated water. Mortality is the rate of deaths per 100,000 man-years (*6 people living for 70 years is 6x70=420 man-years*).

Exposure	Mortality
2.5	147.1
2.6	130.1
3.4	129.9
1.3	113.5
1.6	137.5
3.8	162.3
11.6	207.5
6.4	177.9
8.3	210.3

    a) Make a graph of the data (use graph paper, label completely, and choose the correct axis for the independent(x) and dependent(y) variables).
    b) Find the slope between the lowest and highest exposure counties, accurate to 2 decimal places.
    c) Add a trend line to the graph.
    d) Choose two representative points from your trend line and find the slope. Explain the meaning of the slope in context.
    e) Estimate the y-intercept and explain its meaning in context.

5. The winning Olympic discus throw is recorded through the 1900's rounded to the nearest foot. It is interesting to note that 1916 and 1940 are missing due to the first and second world wars.

    a) Make a graph of the data large enough to include 1992 (use graph paper and label).
    b) Find the slope between 1900 and 1984, accurate to 1 decimal place.
    c) Add a trend line to the graph.
    d) Use your trend line to predict the winning throw in 1992.
    e) Choose two representative points from your trend line and find the slope. Explain the meaning of the slope in context.

Year	Throw (ft)
1900	118
1904	129
1908	134
1912	148
1920	147
1924	151
1928	155
1932	162
1936	166
1948	173
1952	181
1956	185
1960	194
1964	200
1968	213
1972	211
1976	221
1980	219
1984	219

6. Consider the price list for rectangular tarps of different sizes.

    a) Find the area of each tarp in the chart (area of a rectangle is length x width).
    b) Make a graph of the area -vs- price data large enough to include the largest area (use graph paper and label).
    c) Find the slope between the $5 and $15 tarps.
    d) Add a trend line to the graph.
    e) Use your trend line to find a price for the 20'x24' tarp.
    f) Choose two representative points from your trend line and find the slope. Explain the meaning of the slope in context.

Dimensions	Area	Price
5' x 7'		$4
6' x 8'		$5
8' x 10'		$8
10' x 12'		$11
12' x 14'		$15
16' x 20'		$26
20' x 24'		

7. Download speeds for data are increasing with technological advances. The speeds are recorded in the table for each month in 2009.

Month	Speed (Kb/s)
1	7028
2	7056
3	7278
4	7866
5	8188
6	8265
7	8355
8	8529
9	8694
10	8844
11	9183
12	9332

   a) Make a graph of the data large enough to include 10,000 Kb/sec. (use graph paper and label)
   b) Find the slope between months 2 and 11, accurate to 1 decimal place.
   c) Add a trend line to the graph.
   d) Use your trend to find the month when the speed will reach 10,000 Kb/s.
   e) Choose two representative points from your trend line and find the slope. Explain the meaning of the slope in context.

8. Upload speeds for data are increasing with technological advances. The speeds are recorded in the table for each year.

Year	Speed (Kb/s)
2008	1029
2009	1728
2010	2207
2011	2705

   a) Make a graph of the data large enough to include 2015 (use graph paper and label).
   b) Find the slope between years 2008 and 2011, accurate to 2 decimal places.
   c) Add a trend line to the graph.
   d) Use your trend line to make a prediction for the upload speed in 2015.

9. Smoking rates for adults have declined over the last couple of decades.

Year	Percent
1990	25.5
1993	25.0
1995	24.7
1997	24.7
1999	23.5
2001	22.8
2002	22.5
2003	21.6
2004	20.9
2005	20.9
2006	20.8
2007	19.8
2008	20.6
2009	20.6
2010	19.3
2011	18.9

   a) Make a graph of the data large enough to include 2020 (use graph paper and label).
   b) Find the slope between years 1990 and 2010.
   c) Add a trend line to the graph.
   d) Choose two representative points from your trend line and find the slope. Explain the meaning of the slope in context.
   e) Use your trend line to make a prediction for the percent of smokers you might expect in 2020.

10. Health costs are steadily rising in the United States.

    a) Make a graph of the data large enough to include 2020 (use graph paper and label).
    b) Find the slope between years 2000 and 2008.
    c) Add a trend line to the graph.
    d) Choose two representative points from your trend line and find the slope. Explain the meaning of the slope in context.
    e) Use your trend line to make a prediction for the health expenditure you might expect per person in 2020.

Year	Expenditure
1995	$3,748
1996	$3,900
1997	$4,055
1998	$4,236
1999	$4,450
2000	$4,703
2001	$5,052
2002	$5,453
2003	$5,989
2004	$6,349
2005	$6,728
2006	$7,107
2007	$7,482
2008	$7,760
2009	$7,990
2010	$8,233
2011	$8,608

11. Smoking rates for students are shown in the table.

    a) Make a graph of the data large enough to include 2016 (use graph paper and label).
    b) Find the slope between years 2003 and 2011.
    c) Add a trend line to the graph.
    d) Choose two representative points from your trend line and find the slope. Explain the meaning of the slope in context.
    e) Use your trend line to make a prediction for the percent of smokers you might expect in 2016.

Year	Percent
1997	36.4
1999	34.8
2001	28.5
2003	21.9
2005	23.0
2007	20.0
2009	19.5
2011	18.1

Section 4.2: Finding Linear Equations

Rene Descartes (1596-1650) is famous for discovering the connection between geometry and algebra. There is an **algebraic equation** that can be found to match the **geometric line** we see in a graph. Indeed there is an equation to match virtually any geometric shape imaginable.

To discover this simple connection, let's start with the equation $y = \frac{2}{3}x - 4$. Solutions to this equation are ordered pairs since there are two variables. The dependent variable is y since it depends on the choice we make for x. We are free to choose any number we like for x.

x	y
3	-2
6	0
9	2
0	-4
-3	-6
-6	-8

It will be easy if we pick numbers for x that are multiples of 3 since we are to multiply them by $\frac{2}{3}$. The order of operations dictates that we first multiply our choice for x by $\frac{2}{3}$ then subtract 4. It is very important that you are confident with this process, review the order of operations if necessary. The table shows a partial list of ordered pairs.

The graph of these ordered pairs is shown below. The algebraic equation produces a geometric line.

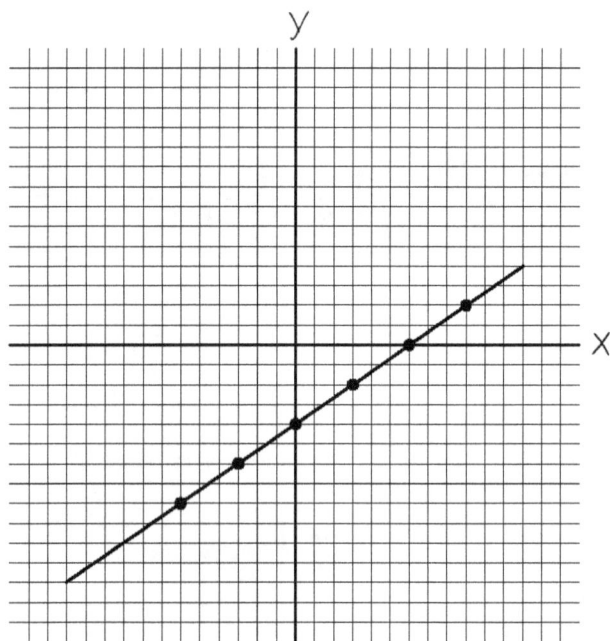

The center of the graph paper (0,0) is named the **origin**.

The **x-coordinate** is located moving right of the origin for a positive x, and left of the origin for a negative x.

The **y-coordinate** is located moving up from the origin for a positive y, and down from the origin for a negative y.

Notice that although 6 ordered pairs are enough to see the pattern, in fact there are infinitely many solutions to the equation since x can be any real number. It is important to understand that the line fills in naturally as you graph more ordered pairs.

The 2-dimensional graph for solutions to 2-variable equations is named the **Cartesian coordinate system** in honor of Rene Descartes (pronounced "day-cart"); a handsome fellow with flowing locks and ahead of his time with a stylish moustache.

Now compare the equation $y = \frac{2}{3}x - 4$ with its graph. It turns out that there is a very simple connection between them. Do you see it?

Hint 1: Find the slope of the line.

Hint 2: Notice where it crosses the y-axis.

The slope is $\frac{2}{3}$ and it crosses the y-axis at -4.

This is more than a fortunate coincidence.

Slope is a rate of change, and the y-intercept occurs where x = 0.

An electric bill is typically the sum of a monthly service fee and a usage fee that is based on the cost of the electricity per kilowatt-hour (KWH). For example, at 5¢/KWH with a $7 monthly service fee, the cost (C) = $\frac{5 \text{ cents}}{KWH}$ K + 7 ... or

C = .05K + 7 where K is the number of KWH used. 5¢/KWH is the rate and $7 is the cost if K = 0.

The equation of a line is said to be in **slope-intercept form** if it is written in the form y = mx + b. The number in place of m = slope and the number in place of b = y-intercept.

Had the English been the first to discover this relationship we might have chosen s for the slope and i for the intercept, but the French beat us to it.

> **Equation of a line:**
> **y = mx + b**
> **m** = slope of the line
> **b** = y-intercept of the line

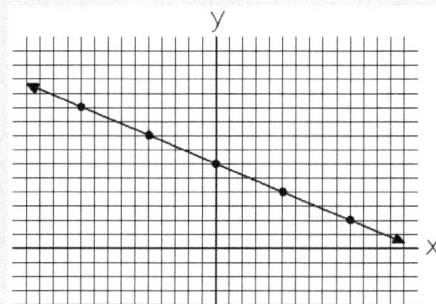

Consider the following example:

Example 4.2.1: Finding the Equation for a graph

Find the equation of the line shown on the graph. Assume each square to be 1 unit.

Solution:

We need two pieces of information from the graph ... slope and y-intercept:

1. y-intercept = 6 since the graph crosses the y-axis at (0,6)

2. slope = $-\frac{2}{5}$, counting the rise and run between any two points on the graph, (reduce the fraction if necessary).

Final Answer: $y = -\frac{2}{5}x + 6$

Note: It is easy to check your work since the points on the graph should "work" in the equation. Try the point (-10,10): $10 = -\frac{2}{5}(-10) + 6 = \frac{20}{5} + 6 = 4 + 6 = 10$ ... it works!

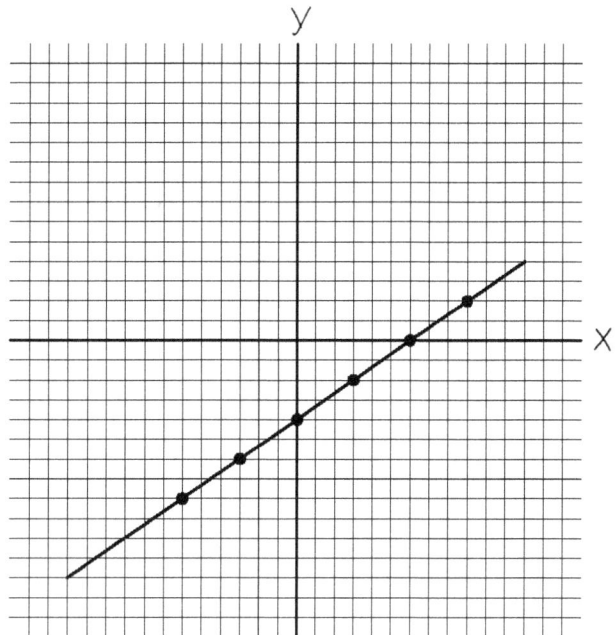

Consider a practical example involving a rental car:

Example 4.2.2: Finding the Equation from an Ordered Pair

Use the chart to find an equation to model the different costs (C) for a rental car based on the miles (M) it is driven.

miles driven	cost
100	$57
200	$82
300	$107
400	$132
500	$157

Solution:

The graph of the ordered pairs in the table reveals that the relationship is linear.

$$\text{slope} = \frac{157-57}{500-100} = \frac{100}{400} = \frac{1}{4}$$

We could have chosen any two points to arrive at this slope since the graph is perfectly linear.

So far we have $y = \frac{1}{4}x + b$

From the graph we might guess that the y-intercept is about 40 but we can do better than guess since the ordered pairs must "work" in the equation. Randomly picking the ordered pair (100,57) from the table, and substituting the values in the equation we get:

$57 = \frac{1}{4} \cdot 100 + b$

$57 = 25 + b$

$32 = b$

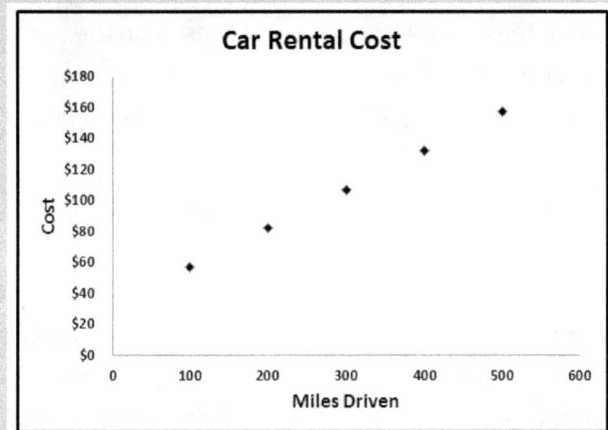

Car Rental Cost

Final Answer:

$y = \frac{1}{4}x + 32$ ... or better $C = \frac{1}{4}M + 32$ ... we now have a formula to find the cost for any number of miles driven.

✎ **Note:** The slope of $\frac{1}{4}$ indicates that the cost of driving the car is $1 for every 4 miles driven or 25¢/mile. Remember that slope is $\frac{rise}{run}$ and the units for the rise (y values) are in dollars and units for the run (x values) are in miles.

Reconsider the practical example involving car value where the points do not line up perfectly:

Example 4.2.3: Finding the Equation for a Line

Use the chart to find an equation to model the value for a car purchased in 1996 based on its age, accurate to the nearest whole number.

Consider 1996 to be year zero. *This is not necessary but it will make the process of finding b and its value much more reasonable.*

year	car value
1998	$27,500
2001	$22,200
2005	$15,700
2007	$12,800
2013	$2,800

Solution:

The graph reveals that the relationship is close to, though not perfectly, linear. Choose 2 points to calculate the slope that accurately represent the trend line. The first and the last seem a reasonable choice here based on the graph.

$$\text{slope} = \frac{27500 - 2800}{2 - 17} = \frac{24700}{-15} \approx -1646.66$$

VERY IMPORTANT ... do not round the slope to the nearest whole number too soon so that b *can be as accurate as possible. Storing the number in the calculator may be helpful in minimizing round off errors and saving time.*

As noted in section 1.1 different points will yield slightly different slopes here since this relationship is not perfectly linear.

So far we have y = -1646.66x + b

From the graph we might guess that the y-intercept is about 30,000 but again we can do better since the ordered pairs must "work" in the equation. Let's pick (17,2800).

2800 ≈ -1646.66•17 + b

2800 ≈ -27993.22 + b

30793.22 ≈ b   a reasonable result based on the graph!

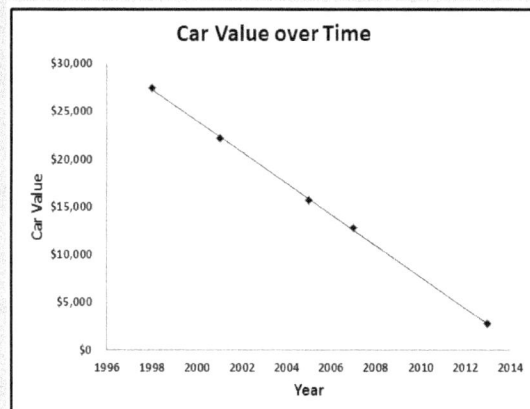

Car Value over Time

Final Answer:

y = -1647x + 30793 ... or better V = -1647A + 30793 (where V = value and A = age after 1996).

Note: The equation indicates that the car was worth about $30,800 in 1996 (year 0). The x-intercept indicates the year the car will be worth nothing and we can now calculate it exactly by plugging in 0 for V then solving for A. Businesses establish depreciation schedules for their assets for tax purposes based on the fact that tools and equipment lose value over time.

Section 4.2: Problem Set

1. The Kilowatt-hours (KWH) of electricity a home uses each month are dramatically affected by the temperature difference inside versus outside (measured in degree days).

   Round slopes and y-intercepts to 2 decimal places
   a) Find the equation of the line passing through the lowest and highest degree day.
   b) Find the equation of the line passing through the 386 and 62 degrees days.
   c) Find the equation of the line through two representative points from the trend line. The points do not have to be from the data.
   d) Comment on the similarities and differences in the equations you found in a – c.

Heating Degree Days (HDD)	KWH Used
714	1799
469	1269
386	1160
266	860
133	804
62	696
7	677
5	666
87	736
354	1071
561	1602
968	2060

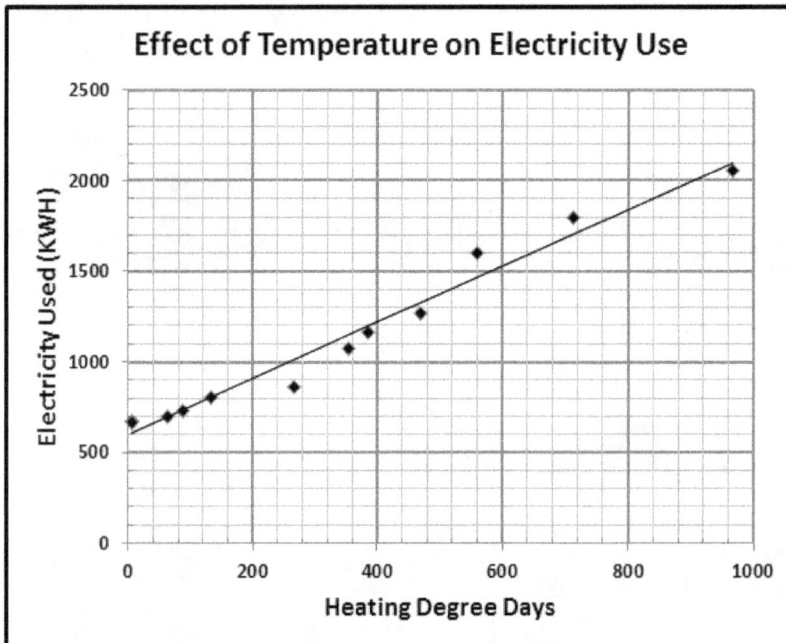

Effect of Temperature on Electricity Use

2. Tree diameter and volume are related and of interest to foresters.

Round slopes and y-intercepts to 2 decimal places

a) Find the equation of the line passing through diameters 7.2 cm and 8.0 cm.

b) Find the equation of the line passing through diameters 8.2 cm and 9.8 cm.

c) Find the equation of the line through two representative points from the trend line.

d) Comment on the similarities and differences in the equations you found in a – c.

Diameter (cm)	Volume (m³)
7.2	143
7.4	157
7.5	168
7.7	178
7.8	189
8.0	199
8.2	212
8.5	221
8.7	237
8.8	248
9.0	261
9.3	272
9.4	278
9.6	288
9.8	299
10.1	304
10.3	318
10.4	327
10.6	340
10.9	349

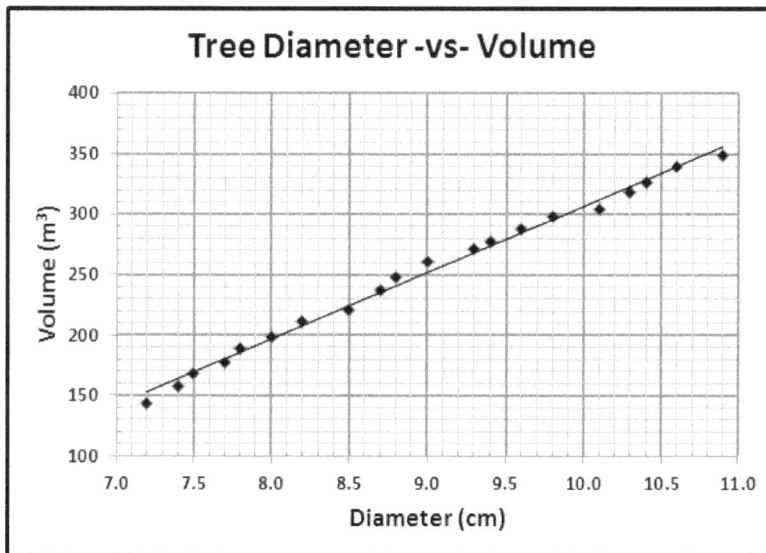

Tree Diameter -vs- Volume

3. A tree's age and its volume are of interest to foresters.

   Round slopes and y-intercepts to 2 decimal places
   a) Find the equation of the line passing through ages 22 and 26 years.
   b) Find the equation of the line passing through ages 30 and 38 years.
   c) Find the equation of the line through two representative points from the trend line.
   d) Comment on the similarities and differences in the equations you found in a – c.

Age	Volume (m^3)
21	143
22	157
23	168
24	178
25	189
26	199
27	212
28	221
29	237
30	248
31	261
32	272
33	278
34	288
35	299
36	304
37	318
38	327
39	340
40	349

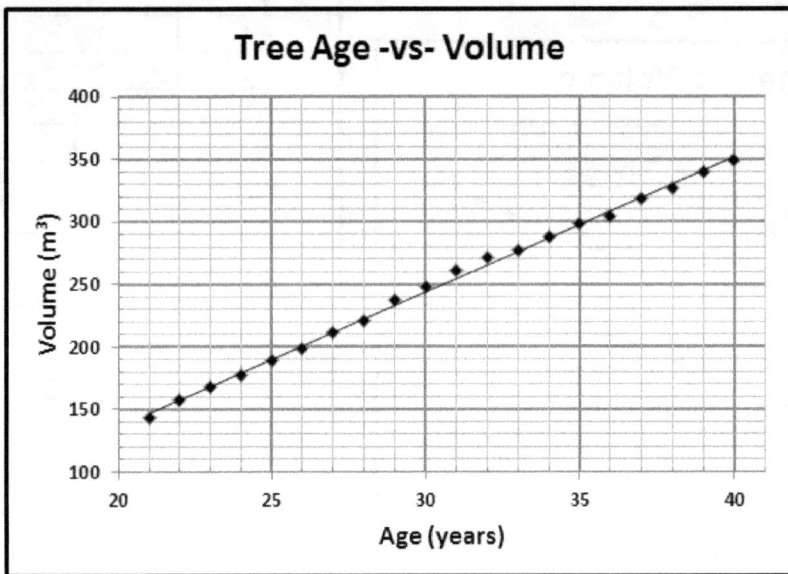

Tree Age -vs- Volume

4. A study conducted in the early 1960's in 9 counties in northern Oregon and southern Washington related exposure to the Hanford nuclear plant and mortality rate. Exposure is an index number assigned to a county based on its population and proximity to the contaminated water. Mortality is the rate of deaths per 100,000 man-years of life.

Exposure	Mortality
2.5	147.1
2.6	130.1
3.4	129.9
1.3	113.5
1.6	137.5
3.8	162.3
11.6	207.5
6.4	177.9
8.3	210.3

Round slopes and y-intercepts to 2 decimal places

a) Find the equation of the line passing through the lowest and highest exposure counties.

b) Find the equation of the line passing through the 3.4 and 8.3 exposure counties.

c) Find the equation of the line through two representative points from the trend line. The points do not have to be from the data and should not be in this case since the line does not pass through any of them very closely.

d) Comment on the similarities and differences in the equations you found in a – c.

Cancer Mortality Near Hanford Reactor

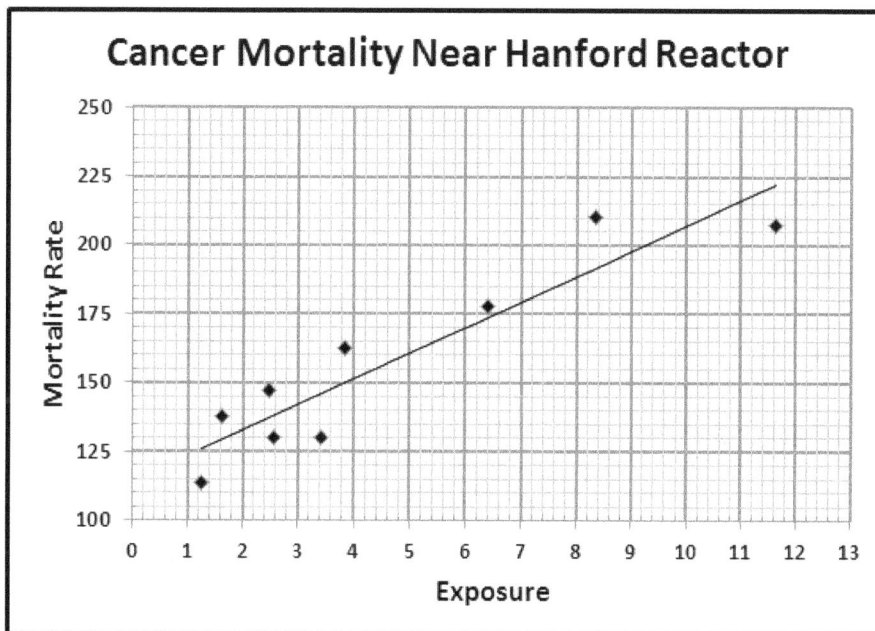

5. The winning Olympic discus throw is recorded through the 1900's rounded to the nearest foot. Consider 1900 to be year 0 for your calculations.

   Round slopes and y-intercepts to 2 decimal places
   a) Find the equation of the line passing through the earliest and latest dates.
   b) Find the equation of the line passing through the years 1912 & 1948.
   c) Find the equation of the line through two representative points from the trend line.
   d) Comment on the similarities and differences in the equations you found in a – c.

Year	Throw (ft)
1900	118
1904	129
1908	134
1912	148
1920	147
1924	151
1928	155
1932	162
1936	166
1948	173
1952	181
1956	185
1960	194
1964	200
1968	213
1972	211
1976	221
1980	219
1984	219

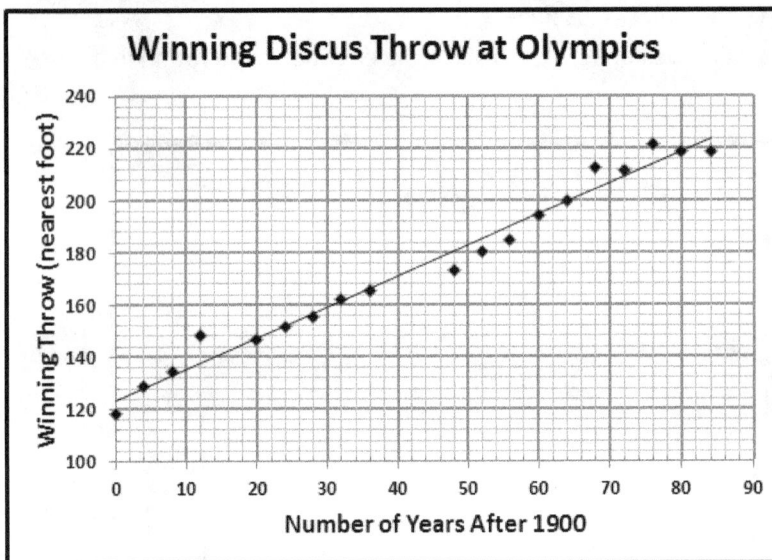

Winning Discus Throw at Olympics

Winning Throw (nearest foot) vs. Number of Years After 1900

6. Consider the price list for rectangular tarps of different sizes.

   Round slopes and y-intercepts to 3 decimal places
   a) Find the area of each tarp in the chart (area of a rectangle is length times width).
   b) Use the $5 and $15 tarps to find a linear equation to model the data using area as the x variable and price as the y variable.
   c) Use your equation to determine the correct price for the 2 larger tarps in the table to the nearest dollar.

Dimensions	Area	Price
5' x 7'		$4
6' x 8'		$5
8' x 10'		$8
10' x 12'		$11
12' x 14'		$15
16' x 20'		$26
20' x 24'		
24' x 30'		

7. Download speeds for data are increasing with technological advances. The speeds are recorded in the table for each month in 2009.

Round slopes and y-intercepts to 2 decimal places
a) Find the equation of the line passing through months 2 & 4.
b) Find the equation of the line passing through months 5 & 7.
c) Find the equation of the line through two representative points from the trend line.
d) Comment on the similarities and differences in the equations you found in a – c.

Month	Speed (Kb/s)
1	7028
2	7056
3	7278
4	7866
5	8188
6	8265
7	8355
8	8529
9	8694
10	8844
11	9183
12	9332

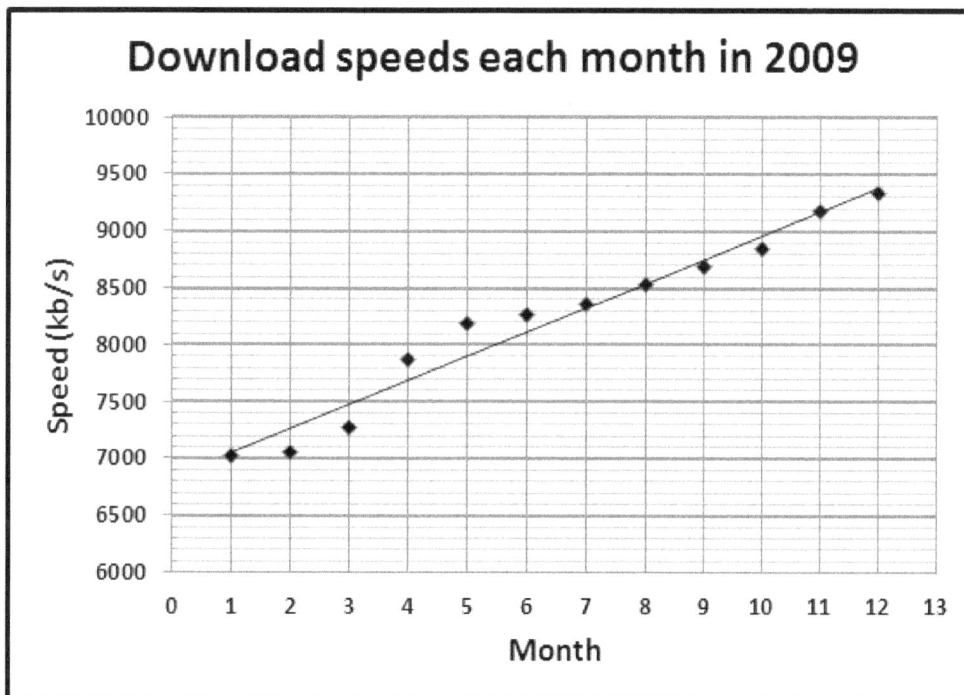

Download speeds each month in 2009

8. Upload speeds for data are increasing with technological advances. Consider 2000 as year 0.

Year	Speed (Kb/s)
2008	1029
2009	1728
2010	2207
2011	2705

Round slopes and y-intercepts to 2 decimal places
a) Find the equation of the line passing through 2008 & 2011.
b) Find the equation of the line passing through years 2009 & 2011.
c) Find the equation of the line through two representative points from the trend line.

Upload Speeds Each Year

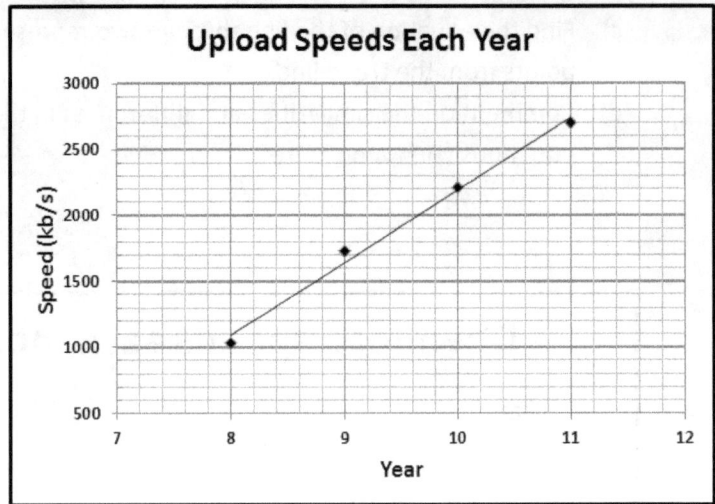

9. Smoking rates for adults have been declining over the last couple of decades. Consider 1990 to be year 0.

Round slopes and y-intercepts to 2 decimal places
a) Find the equation of the line passing through years 2002 & 2009.
b) Find the equation of the line passing through years 1997 & 1999.
c) Find the equation of the line through two representative points from the trend line.
d) Comment on the similarities and differences in the equations you found in a – c.

Year	Percent
1990	25.5
1993	25.0
1995	24.7
1997	24.7
1999	23.5
2001	22.8
2002	22.5
2003	21.6
2004	20.9
2005	20.9
2006	20.8
2007	19.8
2008	20.6
2009	20.6
2010	19.3
2011	18.9

Percent of Adult Smokers

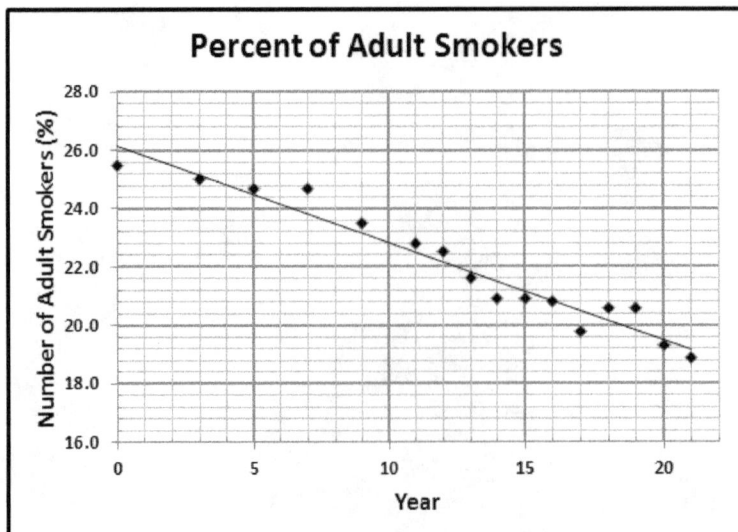

8. Upload speeds for data are increasing with technological advances. Consider 2000 as year 0.

Year	Speed (Kb/s)
2008	1029
2009	1728
2010	2207
2011	2705

   Round slopes and y-intercepts to 2 decimal places
   a) Find the equation of the line passing through 2008 & 2011.
   b) Find the equation of the line passing through years 2009 & 2011.
   c) Find the equation of the line through two representative points from the trend line.

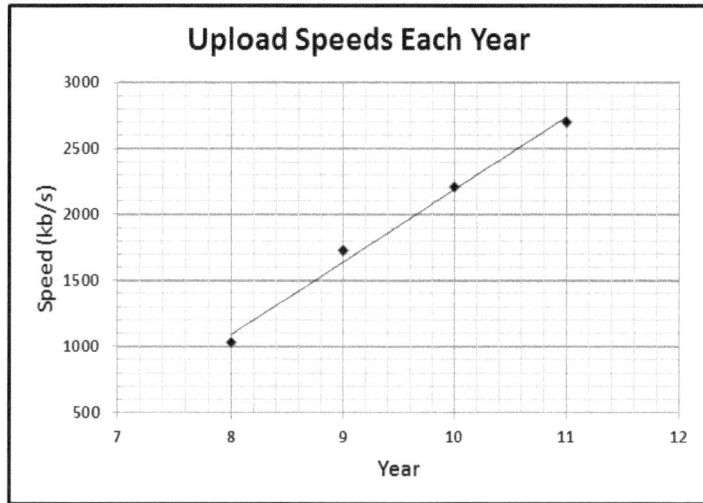

Upload Speeds Each Year

9. Smoking rates for adults have been declining over the last couple of decades. Consider 1990 to be year 0.

Year	Percent
1990	25.5
1993	25.0
1995	24.7
1997	24.7
1999	23.5
2001	22.8
2002	22.5
2003	21.6
2004	20.9
2005	20.9
2006	20.8
2007	19.8
2008	20.6
2009	20.6
2010	19.3
2011	18.9

   Round slopes and y-intercepts to 2 decimal places
   a) Find the equation of the line passing through years 2002 & 2009.
   b) Find the equation of the line passing through years 1997 & 1999.
   c) Find the equation of the line through two representative points from the trend line.
   d) Comment on the similarities and differences in the equations you found in a – c.

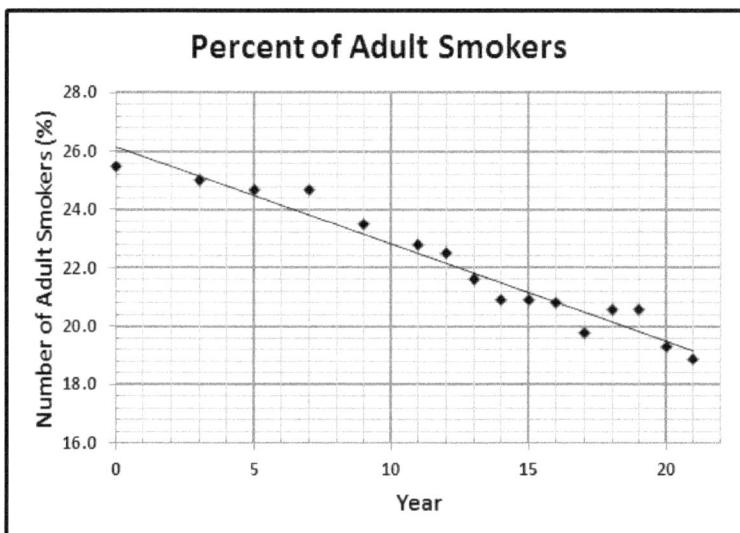

Percent of Adult Smokers

10. Health costs are steadily rising in the United States. Consider 1990 as year 0.

Year	Expenditure
1995	$3,748
1996	$3,900
1997	$4,055
1998	$4,236
1999	$4,450
2000	$4,703
2001	$5,052
2002	$5,453
2003	$5,989
2004	$6,349
2005	$6,728
2006	$7,107
2007	$7,482
2008	$7,760
2009	$7,990
2010	$8,233
2011	$8,608

Round slopes and y-intercepts to 2 decimal places

a) Find the equation of the line passing through years 1995 & 1999.

b) Find the equation of the line passing through years 2000 & 2003.

c) Find the equation of the line through two representative points from the trend line.

d) Comment on the similarities and differences in the equations you found in a – c.

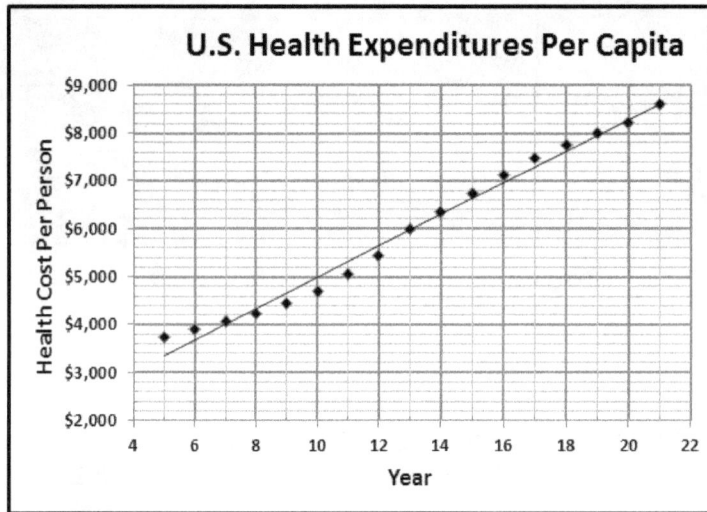

U.S. Health Expenditures Per Capita

11. Smoking rates for students have been declining linearly in recent years as shown in the graph. Consider 1990 as year 0.

Year	Percent
1997	36.4
1999	34.8
2001	28.5
2003	21.9
2005	23.0
2007	20.0
2009	19.5
2011	18.1

Round slopes and y-intercepts to 2 decimal places

a) Find the equation of the line passing through years 1999 & 2001.

b) Find the equation of the line passing through years 2001 & 2005.

c) Find the equation of the line through two representative points from the trend line.

d) Comment on the similarities and differences in the equations you found in a - c.

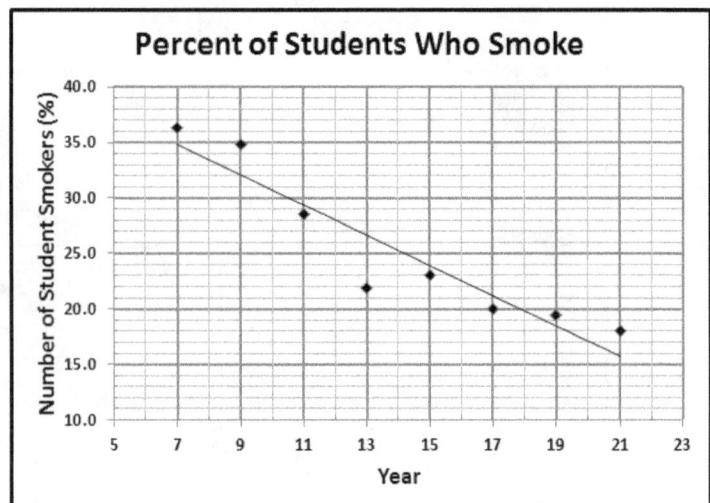

Percent of Students Who Smoke

Appendix:

Common Abbreviations & Symbols:

Metric:

mm = millimeter	cm = centimeter	m = meter	km = kilometer
mL = milliliter	dm = decimeter	L = liter	dL = deciliter
mg = milligram	cg = centigram	g = gram	kg = kilogram

Standard:

" = in = inch	' = ft = foot	yd = yard	mi = mile
gal = gallon	oz = ounce	lb = pound	

Common Conversions:

Length:

Standard Length	Metric Length	Standard to Metric
1 ft = 12 in 1 yd = 3 ft 1 mi = 5280 ft	1 m = 100 cm 1 m = 10 dm 1 m = 1000 mm 1 km = 1000 m	1 in = 2.54 cm 1mi = 1.61 km 3.281 ft = 1 m

Area:

Standard Area	Metric Area	Standard to Metric
$1\ ft^2 = 144\ in^2$ $1\ yd^2 = 9\ ft^2$ $1\ acre = 43,560\ ft^2$ $1\ mi^2 = 640\ acre$	$1\ m^2 = 10,000\ cm^2$ $1\ cm^2 = 100\ mm^2$	$1\ in^2 = 6.452\ cm^2$

Volume:

Standard Volume	Metric Volume	Standard to Metric
$1\ ft^3 = 1728\ in^3$ $1\ yd^3 = 27\ ft^3$ $1\ yd^3 = 46,656\ in^3$ $1\ gal = 231\ in^3$ $1\ ft^3 = 7.48\ gal$ 1 gal water = 8.345 lb	$1\ m^3 = 1,000,000\ cm^3$ $1\ cm^3 = 1000\ mm^3$ $1\ L = 1000\ cm^3$ 1 L = 1000 mL 1 L = 10 dL $1\ mL = 1\ cm^3$	$1\ in^3 = 16.387\ cm^3$ $1\ oz = 29.574\ cm^3$ 1 gal = 3.785 L

Weight:

Standard Weight	Metric Weight	Standard to Metric
1 ton = 2000 lbs 1 lb = 16 oz	1 g = 1000 mg 1 kg = 1000 g	1 ton = 907.2 kg 1 oz = 28.35 g 1 lb = 453.6 g

Plane Figure Geometry Formulas:

Name	Figure	Perimeter/Circumference	Area (A)
Rectangle		$P = 2L + 2W$	$A = LW$
Parallelogram		$P = 2a + 2b$	$A = bh$
Trapezoid		Add all four exterior lengths	$A = \frac{1}{2}h(a+b)$
Triangle		Add all three exterior lengths	$A = \frac{1}{2}bh$
Circle		$C = 2\pi r$ **for a circle, perimeter is renamed circumference since it is the measure of a curve	$A = \pi r^2$ $A = \frac{\pi d^2}{4}$ this formula can be used if the diameter (d) is known instead of the radius
Sector		$L = \frac{\theta}{180}\pi r$ **for a sector, perimeter is renamed arc length	$A = \frac{\theta}{360}\pi r^2$
Ellipse		$C = \pi(a+b)j$ $j = 1 + \frac{1}{4}h + \frac{1}{64}h^2 + \frac{1}{256}h^3 + ...$ $h = \frac{(a-b)^2}{(a+b)^2}$	$A = \pi ab$

Solid Figure Geometry Formulas:

Name	Figure	Surface Area (SA)	Volume (V)
Rectangular Prism		$SA = 2wl + 2hl + 2wh$	$V = lwh$
Triangular Prism		$SA = ab + d(a+b+c)$	$V = \dfrac{1}{2}abd$
Cylinder		$SA = 2\pi r^2 + 2\pi rh$	$V = \pi r^2 h$ $V = \dfrac{\pi d^2 h}{4}$ this formula can be used if the diameter (d) is known instead of the radius
Pyramid		$SA = b^2 + b\sqrt{b^2 + 4h^2}$	$V = \dfrac{1}{3}b^2 h$
Cone		$SA = \pi r^2 + \pi r\sqrt{r^2 + h^2}$	$V = \dfrac{1}{3}\pi r^2 h$
Frustum of a Pyramid		$SA = b^2 + B^2 + (B+b)\sqrt{(B-b)^2 + h^2}$	$V = \dfrac{1}{3}h\left(B^2 + Bb + b^2\right)$
Frustum of a Cone		$SA = \pi r^2 + \pi R^2 + \pi(R+r)\sqrt{(R-r)^2 + h^2}$	$V = \dfrac{1}{3}\pi h\left(R^2 + Rr + r^2\right)$
Sphere		$SA = 4\pi r^2$	$V = \dfrac{4}{3}\pi r^3$

CPSIA information can be obtained
at www.ICGtesting.com
Printed in the USA
BVHW021309260223
659219BV00014B/657